T0109932

Merde

EXCURSIONS IN SCIENTIFIC, CULTURAL, AND SOCIOHISTORICAL COPROLOGY

RALPH A. LEWIN

RANDOM HOUSE
NEW YORK

Copyright © 1999 by Ralph A. Lewin

All rights reserved under International and Pan-American Copyright Conventions. Published in the United States by Random House, Inc., New York, and simultaneously in Canada by Random House of Canada Limited, Toronto.

Library of Congress Cataloging-in-Publication Data
Lewin, Ralph A.
Merde : excursions in scientific, cultural, and sociohistorical coprology / Ralph A. Lewin
p. cm.
Includes bibliographical references and index.
ISBN 978-0-812-99251-9 (hc.)
1. Scatology. I. Title.
GT3055.L48 1999
392—dc21 98-27259

Random House website address: www.atrandom.com

Printed in the United States of America on acid-free paper

146406575

To my mother, who changed my nappies,
and to my father, who occasionally helped

"What cannot be used from the consumed foods and drinks descends into a person's lower intestines, changes itself into excrement . . . and is evacuated by the body."

—Hildegard von Bingen (ca. 1155),
Causae et Curae

PREFACE

The title of this work is a direct quotation of an expression used by the French general Cambronne in 1815, after the defeat of his forces at Waterloo. Victor Hugo proclaimed it to be "perhaps the finest word ever spoken by a Frenchman."

The editor suggested I add a Preface. He asked me to explain why I chose to compile a book on this subject. I suppose it's because of a natural interest in an activity that occupies every one of us for several minutes almost every day throughout our lives.

It has been said that "we are what we eat." Of course, like most aphorisms, this is only partly correct. What we and other animals eat is carefully processed in our bodies, and only the few important components that we need for constructing and operating our corporeal systems are selectively extracted. The residual bulk, often neatly packaged, is more or less regularly expelled. This constitutes the feces, the matter of coprology, and this is what the present book is all about. I have tried to distinguish between coprological subjects, involving the study of feces, and scatological material, which is generally understood to refer to dirty words. I think the former is more interesting.

And now, dear Reader, let me ask you: why did you choose to buy, borrow, or at least glance through it?

CONTENTS

INTRODUCTION

Somewhere along the Alaska pipeline, between Fairbanks and the Arctic Ocean, our guide pointed out some "moose nuggets," in a joking allusion to the gold rush of a century ago. Although we saw few moose (in Europe, elk) in the flesh, we found many of their droppings, in loose clutches nestled among the reindeer mosses. Each batch consisted of a couple of dozen pellets wonderfully uniform in size and shape, like large olives or khaki-colored silkworm cocoons, 3 centimeters long and 1.5 centimeters wide.

Three years later, in the swampy llanos of Venezuela, I saw piles of similar pellets, produced by a very different kind of animal, a capybara. This is a knee-high rodent, quite unrelated to ruminants such as the moose. There were also piles of larger, shapeless feces, which our native Indian guide told us were likewise produced by capybaras, but by sick ones infested with worms. All this set me wondering: why has natural selection favored the normal production of so neat a waste product? And that led me in turn to consider the wider subject of what we might call comparative coprology in general, and begin this survey of what seems to me a fascinating if neglected topic. Although the word "scatology" has been widely employed for this subject, I think that "coprology" is to be preferred; it sounds more respectable.

There is of course a vast store of scatological humor in all national cultures. Not only Mark Twain, but also Mozart, George Washington, Sir Winston Churchill, and Justice Clarence Thomas are known to have indulged in it, though not always with humorous intent. (Who hasn't?) *Clean and Decent* (1960), compiled by the architect L. Wright, is as full of wry humor as it is of scholarship. There is also a fair amount of scatological art—or what passes for art—that has just been extensively surveyed, with abundant illustrations, in *The Art Journal* (Weisberg, 1993). (The article in this issue bears the title "Merdre," as used in the play by Alfred Jarry *Ubu Roi:* the extra "r" was apparently added for emphasis, and not as an attempt at bowdlerization.) One small example is a defecating gargoyle on a cornice of the fifteenth-century town hall at Noyon (northeast of Paris); I doubt whether such works of art would be approved in contemporary architecture today. Since it was well reviewed by *The Art Journal,* this arcane aspect of the subject, too, I shall pass over only lightly.

In the last century, probably the most authoritative work on the subject in English was that of the scholarly Captain John G. Bourke, 3rd Cavalry U.S.A., who compiled *Scatalogic Rites of All Nations, a Dissertation upon the Employment of Excrementitious Remedial Agents in Religion, Therapeutics, Divination, Witchcraft, Love-Philters, etc., in All Parts of the Globe.* (The original publication, by Lowdermilk & Company in Washington, D.C., is marked "Not for general perusal.") Bourke made extensive use of the U.S. Army Medical Museum Library, especially a "learned pamphlet" entitled *Bibliotheca Scatologica* lent to him by Surgeon John S. Billings. He reviewed the international literature on feces, as well as many unrelated subjects, covering ancient and obscure sources back to (and even before) Pliny, and leaning heavily on a book called *Chylologia,* published in Dresden in 1725 by Schurig et al. (which I have not myself seen). Bourke's chapter on "ordure and urine" alone runs to ninety-four pages; that on witchcraft, to another thirty-three, while others less long deal with "insults" and "myths." He has special sections on the dung gods of the ancient Romans, Egyptians, Moabites, and Israelites. He tells us, for instance, about the unfortunate Roman emperor Vitellius, who was pelted by dung before being executed. But those who wish to consult Bourke for further information of this sort should be aware that much of his text is in languages other than modern English, including Latin, Greek, archaic French, German, and Spanish.

More recent times have seen the publication of *End Product* (1977) by the "anthroscatologists" Sabbath and Hall. It is a comprehensive exposition on human excretion, dealing with physical, psychological, and cultural attitudes to the generation, passage, and elimination of feces and other smelly matters. It reviews, with digressions in all directions, many of the intricacies of human and urban plumbing and their relevant terminologies. And it quotes innumerable remedies for this and that, made from dungs of all sorts and provenances. Unlike the compendium of Bourke (to which frequent references are made), its consultation by the general public is not proscribed, but it seems to have been out of print for many years. I have quoted a number of items from this work (as I have from Bourke) without special attributions to the original sources, many of which are not readily accessible. For small children, an introductory booklet on the subject, originally written in Japanese by Taro Gomi, has more recently been published in an English translation, *Everyone Poops* (Brooklyn, New York: Kane/Miller, 1993). And for more specific information on toilets, the reader is referred to a scholarly historical review by the novelist and biographer Wallace Reyburn (1971).

This work is presented as a more or less scientific *aggiornamento.* I've tried to stick fairly closely to the broad subject of fecal matter. (Readers seeking comparable reviews on urine must look elsewhere.)

Merde

1

TERMINOLOGY AND CULTURAL ATTITUDES

The word "science," meaning knowledge, and the word "shit," from the Old English *scitan,* both apparently derive from the same ancient Indo-European root, as does the Greek word from which *scatology* is derived. (The modern Greek expression *Skata sta moutra sou* is not recommended for polite society.) For *cowpats,* however, the Greeks had a special word, *bolita.* The parallel between the German words *unterscheiden* (to distinguish) and *ausscheiden* (to excrete) supports this initially implausible derivation. Shit is one of the few English words that have identical forms in the present tense, past tense, and past participle. (The past tense is also *shit,* not *shat.* A *shat* was a weight unit for silver in ancient Egypt.) Some foreigners learning English have to make a special effort to pronounce the word "sheet" with a long vowel, otherwise their remarks are likely to be misconstrued. (An equally embarrassing mistake may be made by foreigners learning Japanese who fail to use a long vowel for the first syllable of *kuso,* reverie.) Back in 1300, the London street Sherbourne Lane gained the familiar appellation "Shitteborne Lane" because of its exposed sewage. The Greek word, like the related Anglo-Saxon root "scat," has given rise to the formal name for a stinking indole derivative, skatole, one of the major components contributing to the smell of mammalian feces, as well as certain species of evil-smelling ants. It is interesting to note that the Latin singular noun *faex* was rarely used in Roman times (except in reference to the lees of wine), and is apparently never employed in English, where we use only the plural, *faeces,* or as in American spelling, feces. Swahili, likewise, has only the plural word *mavi;* in a logical language, like Esperanto, either the singular, *feko,* or the plural, *fekoj,* is equally acceptable.

In medieval times, dung was often referred to as "gore," but this meaning has now gone out of use. In the King James version of the Bible, the accepted euphemism for defecation is "covering one's feet." This is generally accomplished in what Englishmen refer to as "public conveniences" or "water closets," i.e., W.C.s; in France this term is often contracted to "le water." In America such a place is designated as a "rest room" or "powder room" where one may go to "wash one's hands." Informally it is often referred to as the "john," reminding us that in Elizabethan England it was called the "jakes" (cf. Jacques). In Britain it is commonly called the "loo" (cf. French *l'eau*); in Nigerian trains, look for the "choo."

In the pidgin of the Solomon Islands—where custom dictates that ladies should precede gentlemen in their morning defecations into the waves—the preferred expression is "Mi go haus pek-pek" or "Mi go mambis." The list of such terms is endless, and constantly evolving, but in this book I shall try to avoid unnecessary euphemisms and shall generally hew to the Dunstable way and call a spade a spade. Incidentally, in reading Hebrew scriptures aloud, one is instructed to pronounce neither the Hebrew word for God (*Yahveh,* our Lord) nor that for dung, but should substitute conventional alternatives in order to avoid offending the sensitivities of listeners, if not God.

T. H. White, in *The Sword in the Stone* (1939), referred to droppings of the Questing Beast by the ancient word "fewmets." In his story, they had to be collected in a hunting horn as objective evidence that the quarry had been located. (This was not a pure fictional invention of White's: he must have read somewhere that hunters in Angola used to put dung into their hunting horns for good fortune.) The more usual spelling, for deer droppings, is "fumets," a word derived through the Latin *fimus,* from the Greek *thyma,* which meant "an offering" and was probably used as a sort of joke.

It seems a shame that the good name of bullshit, a potentially useful product, should have been debased in recent parlance to signify worthless or misleading statements. The use of the word "poppycock," which in American speech has much the same meaning, is considered more genteel, though it is derived from a comparable Dutch dialect expression, *pappekak,* soft dung.

(Packages of popcorn currently displayed in American supermarkets under this name are presumably prepared and sold by people unaware of its derivation.)

The word "shit" has been widely used to express various forms of disappointment, dismay, or disgust. No less a personage than the Prince of Wales, before he became King Edward VIII, recorded in his diary on a particularly black day "I feel like a filthy little shit." Albert Einstein is quoted as saying (in German or English, I'm not sure), "We remain students as long as we live, and don't give a shit for the World." Robert Strauss, U.S. principal trade negotiator, said, "I'm used to bullshit because I'm from Texas, and it's all over the fields down there. . . . You have to learn how to keep from stepping in it." When in 1994 the American congressman Dan Rostenkowski was accused of financial irregularities, his supporters asserted that the charges against him were "chickenshit," presumably less weighty than the bovine variety. In the same way, Saint Paul vilified all sorts of worldly matters by the word "stercŏro," which has much the same meaning. (I am not sure that the motto *Stercus bubulus omnia vincit* is of genuine classical antiquity, but it certainly seems authentic.) More recently Aleksandr Solzhenitsyn denounced Western popular culture in the same terms (but in Russian, of course). And when Governor Clinton was told that presidential preelection polls indicated a reduction in his popular support, he said despondently, "I'm dropping like a turd in a well." A Newtonian apple would fall equally fast, naturally, but the fecal metaphor may have been chosen as more illustrative of Clinton's disappointment.

It might be preferable to use the word "crap," which in its original meaning seems to have referred specifically to such useless materials as chaff. The word "crapper" entered into U.S. slang only as recently as the 1920s or 1930s. (For details about a plumber eponymously called Thomas Crapper, see Chapter 8.) Colloquial French is not much less unsympathetic to this valuable commodity than English. "*La loi d'emmerdement maximum*" is a Gallic restatement of Murphy's Law. *Merde du Diable* is a term that has been used for the drug asafetida. Word meanings drift to and fro. A "stool sample" refers to a fecal specimen produced while sitting on a closestool, which has an appropriate hole in the seat. Thus the stool is not the product but the means whereby it is obtained. (The word "tax" has undergone roughly the same evolutionary change.) I recently came across the expression "a call to stool," and I think I can guess its meaning although I cannot find it in either an Oxford or a Webster's dictionary. Meanings change in odd ways. When many of us in Britain were children, we were "evacuated." This didn't mean that we were in any way purged; it indicated that we were involved in the evacuation of London, in order to avoid the German bombing anticipated at the outbreak of World War II.

Perhaps in recognition of its human value, the Chinese word for feces, *da-bien,* the "big convenience" (as distinct from the more fluid "small convenience"), is not generally used as a vulgar expletive, a feature in which Chinese language customs differ from those of Western Europe and the Americas. Nevertheless, the expletive use of this term seems natural. Even among our nonhuman relatives, chimpanzees that have been taught sign language have been observed to indicate feelings of frustration or anger by giving the sign that they have learned to signify feces.

It has been said that the Maori language boasts of some thirty-five words for dung, though it is questionable whether they all have clearly distinct meanings. In Alsace-Lorraine, horse manure is sometimes poetically called *Pferd-merde,* thereby combining Teutonic with Gallic elements. (For a more exhaustive review of this subject the reader is referred to a recent book on *Dirty Words* by A. Arango, 1989.)

If we are to believe assertions in a thesis by Dundes (1984), what amounts to almost an obsession with feces, as bases for obscenities, insults, and jokes, has characterized much of Germanic culture. (In German, farmyard manure is known by the rather nebulous word *Mist.*) Sigmund Freud (who suffered from constipation for much of his life) contended that an interest in excretion is a trait naturally developed in childhood: feces are, after all, the first things that children are encouraged to produce by themselves. One of his patients, Sabina Spielrein—later to become an even closer acquaintance of Freud's—was prone to thinking about feces at mealtimes (which presumably had some effect on her appetite). Hitler reportedly had a kind of fetish about feces. A slim novel by Patrick Süskind, reviewing his protagonist's feelings when confronted with the need to defecate while being watched by a pigeon, originally published in German, later appeared in an English translation (Süskind, 1988). Jokes and illustrations about defecation feature prominently in Steinbart's amusing work on doctor-patient relations (1970), where several pages are devoted to clysters, a purging device popular in earlier centuries. And in further support of Dundes's contention, it may be mentioned that, in a book published in 1840, the eminent Justus von Liebig indicated "that society needs manure more than mathematics."

Of course, Germans are by no means unique in this respect. François Rabelais, and Queen Marguerite of Navarre who wrote *The Heptameron* shortly after Columbus's rediscovery of America, were French. An Englishman, Bishop John Fisher, who rose to be chancellor of Cambridge University in the early 1500s and was more recently canonized, despondently described the human body as "a sachell full of dung." A critic disprized one of the less familiar productions of the Cambridge University Opera Society as a "heap of mastodon droppings," though neither he nor anyone living today could have seen such things. Among the countless other figurative uses of coprological terms we might also quote from *Pilgrim's Progress*, where the pious John Bunyan refers to a purge made from the body and blood of Christ, although we must assume that this should be taken spiritually rather than literally. In polite Anglo-Saxon society today, public discussions of human feces are rare. However, it is reported that the eminent pathologist H. G. Burkitt used to hold forth on the subject of

African native turds, in a loud voice, in genteel restaurants generally unused to pronouncements on such esoteric subjects. A favorite story of his related how certain villagers, on being asked to supply stool samples, proudly lined up outside his makeshift office with their offerings neatly assembled or coiled on paper plates. But he didn't refer to them as "stool samples."

Geographers should be careful to spell the Brazilian town name Caçapava with a cedilla under the second "c," so that it should not be misread to mean "turkey shit."

2

PHYSICAL FEATURES: SHAPES AND SIZES

The digestive systems of the simplest invertebrates—for example, jellyfish—lack a continuous throughput passage: excretion of undigested residues has to be effected by regurgitation through the mouth. (As recently as the nineteenth century, some lizards such as Gila monsters were wrongly believed to lack a cloaca and to excrete solely by vomiting.) Differentiation of a distinct anus was a great step forward in evolution. Many of the more advanced invertebrates that burrow in mud or sand produce characteristic heaps of excreted material, like the familiar worm casts on our lawns and golf courses, because the ratio of mineral matter to digestible substances is so high. The feces of many sea cucumbers are similar, and may be as long as the animal itself; those of *Psychropotes* are helical even before they are expelled. (For illustrations and further information on the shapes, sizes, and dispositions of fecal pellets produced by sea cucumbers and other invertebrates on the sea bottom, see Chapter 5, on "Excrement," in Heezen and Hollister, 1971.) The feces of some enteropneust worms too are usually spiral, and thus quite characteristic, a useful thing to know if you happen to be a marine biologist hunting for such creatures. Many sharks and their allies likewise produce spiral feces. They are perhaps the same species as those that produce spiral egg cases, since a common orifice is involved in both processes. Some coprolites (fossilized feces), including ones of certain Irish fish from the Lower Paleozoic era, are also spiral, providing information on the conformation of the creatures' cloacae, which are normally not fossilized.

Olaus Murie's *Field Guide to Animal Tracks* (1982) and the more recent field guide by Halfpenny and Biesiot (1986) are well illustrated with sketches and photographs of the scats of North American mammals, from shrews to buffalo, and there are comparable handbooks for animals in Europe and Africa. Murie's book illustrates variations in sizes and shapes of rodent pellets and ungulate turds, including several different kinds of deer droppings produced at different seasons, and gives instructions on how to prepare and preserve a scat collection. (Why not be the first on your block to compile one?) Halfpenny and Biesiot include a fourteen-page section on the scientific study of coprology, which, they claim, though "once solely the realm of the naturalist, has grown into a valuable tool for ecological research." They have also presented us with a useful two-page tabulation of the general features of

North American mammal feces, and they too give valuable instructions on how best to collect and handle, photograph, dry, varnish (if necessary, to retain their integrity), and store them (they suggest, understandably, in an outside garage) in a personal collection or for exchange. There are some artistic colored photographs of droppings of European animals in Bang and Dahlstrom's book on tracks and signs (1974): the one of fallow-deer fumets is quite beautiful, at least to those with an eye for such things. Many scats resemble the food pellets that one can buy for pets such as rabbits, which is not surprising since a rectum, or an intestinal section that precedes it, can be regarded as a natural pelleting device. In long grass, some droppings may look a bit like small animals: indeed, a coprophilous (and perhaps shortsighted) cane toad has been filmed in Australia earnestly trying to copulate with a horse dropping.

A delightful booklet, with abundant colored illustrations, data, and anecdotes, was published in 1991 by Hansard and Silver. It purports to serve as a field guide to the flattened fecal matters—referred to there as "splats" or "splays"—that some forty North American birds deposit on the windshields of cars. The authors have also included a short note on bat splats. Although this booklet is certainly among the most entertaining of all the works cited here, one is led to question the authenticity of the information it contains. (Healthy geese, I know, do not make splattery splotches like those they have illustrated, and many other entries, though perhaps more plausible, are equally suspect.) However, as an amusing item for coprophile-bibliophiles the book is to be strongly recommended.

Turds may be single (hares), composite with faceted segments (sheep), or clumped (rabbits, pigs, and pronghorns). They may be more or less spherical (hares and their ilk; to the cognoscenti the droppings are known as "crottels"); flattened or even concave at one end, where a harder pellet has pressed against a softer one before deposition (elk, fallow deer); oblong, cigar-shaped, cylindrical (goats), or even coiled (foxes); terete, grooved longitudinally (coypu) or transversely (zebras); segmented or joined in chains like short rosaries (porcupines, ground squirrels); blunt; pointed (duiker antelopes); tapered or tailed at the back end (prairie dogs, toads) or at both ends (weasels); dry or squishy, all according to species, age, season, and diet. If they are deposited in special latrine heaps, like those of voles, they may fuse into shapeless masses. If they are deposited in water, as are sometimes those of bobcats, beavers, and raccoons, they may disintegrate altogether, though those of beavers may remain coherent and float because of their content of undigested wood. Those of jumping hares may even vary according to sex. Since mammalian guts, like racetracks, have a straight section at the end (hence the word "rectum"), most long and coherent turds (notably those of many carnivores) are correspondingly straight when they are expelled, though they may fall apart into segments (such as those of cats).

On a dry diet, the final turds of sheep remain compact and discrete, whereas in sheep on more succulent diets, such as wet grass, the individual shitlets become more or less compacted at the end of the rectum, and are eventually discharged as morular lumps—as any observant shepherd could tell you. As one might expect, most of the pellets are left on sites that have been occupied overnight. Deer fumets tend to resemble those of sheep. Caribou, like moose, are deer, and in Alaska one sees lots of their droppings, which are in shape and color more like raisins (at least, that's what the ones in Alaska looked like to me), and are sometimes glazed with dried mucus. Caribou are smaller than moose admittedly, but this doesn't entirely account for the size difference in their excreta. Possibly each pellet represents a module of food at the end of a sequence of peristaltic waves along the intestines, but in view of the smallness of caribou pellets maybe this assumption is wrong. Llamas, which live on upland grasslands, typically produce pellets no bigger than those of rabbits, despite the considerable difference in the sizes of the animals that drop them. However, within a single species one can often estimate the body size from the dimensions of the pellets. Red deer stags make bigger fumets than their smaller does. Likewise male giraffes, perhaps because they are larger than females, are reported to produce larger feces, although since their droppings fall from a considerable height, they tend to fragment on reaching the ground and thereby obscure the difference.

Desert animals have to conserve water as much as possible, and cannot afford to waste it in excrements. The drier the climate, the drier the vegetation, and thus the drier and harder the scats of vegetarians. Some camel droppings are so dry that they are combustible as soon as dropped. In parts of the Sahara Desert, where there is plenty of sand but few or no stones, camel droppings are sometimes used instead of pebbles as counters for ground games like the European game of Nine Men's Morris. (I do not know how the tokens of the two sides are distinguished.) Some desert-hare pellets, which I found in clusters on the Sinai Desert, were so desiccated, hard, and shiny that at first I thought that they were seeds of an unknown plant. Further to add to my confusion, they were odorless, being too dry even to rot. (These are fecal characteristics that help to make gerbils rather attractive as pets.) The droppings of kangaroos on the semidesert plains of Australia, though larger of course, are just as hard and round.

In general, bigger animals produce bigger turds, and the more they eat, the more they excrete, whereas people who for some reason take in no solid food—for instance, those fed solely by intravenous injection of liquids—produce none at all. The relative amounts of fecal matter produced by animals range from a daily bulk production tens or hundreds of times that of the animal itself, e.g., lugworms that ingest mud, and reveal their presence to anglers by the voluminous casts they leave on mudflats around the coasts. In contrast, the mites that live in hair follicles, and the recently discovered deep-sea vestimentiferan worms, since they eat nothing solid, produce no feces at all (nor do rocking horses, to quote a colloquial Australian expression indicating extreme rarity). Such worms live by symbiosis, depending on the activities of associated sulfur-oxidizing bacteria to produce all their needed nutrients in soluble form.

Snakes like the Guatemalan jumping viper, which eats all too rarely, defecate only every month or two, though when they do so they usually produce a generous amount of fecal matter. On the other hand, a rabbit may excrete more than five hundred pellets a day, and a large panda, on a relatively poor diet of bamboo leaves and shoots consisting largely of indigestible fibrous material, can daily produce a hundred or more. The normal score for a well-fed cow is about twelve, all typically sloppy. Apparently whenever they feel the urge, horses drop a relatively modest 5 kilograms daily, whereas at the other extreme some small mammals produce only a few little pellets weighing no more than 100 milligrams. A large elephant may defecate every few hours, at each event releasing 6 to 30 kilograms of partly digested vegetation. In the Tsavo game park alone, some 1,500 tons of elephant droppings are produced every day. Such masses stay warm for several hours after being dropped, providing a useful clue for trackers. It has been estimated that at some seasons, four thousand tons of wildebeest droppings fertilize the Serengeti daily.

An adult salmon, on a protein-rich diet, produces about a kilogram (dry weight) of feces in a year. (Compare that with the elephant, which excretes almost fifty times as much every day.) At the other end of the scale, microscopic dinoflagellates in Antarctic marine plankton, barely 30 micrometers in diameter, may produce fecal pellets as large as themselves (Buck et al., 1990). (They contain the skeletons of diatoms, which are siliceous and spiky; perhaps for this reason, the pellets are generally surrounded by a special membrane. In this they resemble droppings of baby birds; see Chapter 13.)

The shapes of insect feces are not as well known as those of mammals, which have always been studied by hunters and trappers. But, though on a considerably smaller scale, they are often equally characteristic. The frass kicked out by wood-boring insects such as termites and deathwatch beetles is one of the most evident indications of insect infestation in buildings and furniture. On a quiet summer evening, if you stand under a linden tree in which many looper caterpillars are browsing you may hear the pitter-patter of their droppings falling on the leaves below their suppers. Now look down, and you will see how nicely these pellets are made: short cylinders, each with six parallel grooves on the sides, shaped by the rectal pouches. The form of caterpillar droppings is generally fairly specific; those of the larval viceroy butterfly are particularly neat. Ecologists of woodland trees gain much information by studying the forms and quantities of insect feces that accumulate in traps slung under the branches or on the ground below (Komata et al., 1994). In a temperate woodland during the course of a year, the total weight of caterpillar droppings has been estimated at half a ton per hectare. (The value for earthworm casts produced in the same area could be fifty times as high.)

Among the most remarkable insect feces are those produced by certain large dung beetles (*Pachylomera*) when fed a concentrated extract of sheep dung. If held over a pot, the beetles could be induced to produce yards of excreta, consisting largely of gut lining, every day. Dr. K. M. Rudall of Leeds, whose work I quote here, was so impressed that he "dedicated these most excellent, smooth, uniform, long, continuous tubes to Professor A. Glenn Richards," an eminent entomologist who, I'm sure, must have been delighted to be so honored. Somewhat similar are the elegant, looped fecal strands normally produced by the leaf-mining caterpillars of *Lyonetia* moths, although they are rarely appreciated except by people who wander through mangrove swamps.

3

PHYSICAL FEATURES: COLORS AND TEXTURES

The whiteness of many bird droppings is mainly attributable to crystals of uric acid, produced by a special metabolic process to eliminate excess nitrogen while conserving water. On the dancing grounds (known as "leks") of such birds as ruffs, the feces are soon trampled into characteristic white patches among the moors where they breed, and this of course accelerates their ultimate disintegration. I have noted that each of the round droppings of a dove, nesting in the honeysuckle on my back porch, bears a central blob of white icing, which I take to indicate either that the uric excretion is the last component to leave the bird's cloaca, or else that it is the lightest part of the dropping which trails as it falls. I suspect the former explanation is correct, since I have noted that goose pellets tend to be white at the hind end. Although the phenomenon may be common among birds, I don't recall having noticed it among the droppings in the wood-pigeon cages that I had to clean out daily as part of my first job in Oxford many years ago.

The black-and-white pattern of some bird feces is so characteristic that certain tropical spiders, moths like the North American *Ethmia*, and some kinds of sawfly and butterfly caterpillars (e.g., species of *Papilio*) have evolved body colorations that mimic them. They thereby help to conceal themselves from predators or to attract coprophilous insects, such as *Heliconius* butterflies, on which the spiders may prey. There are rove beetles in Central America which not only disguise themselves as bird droppings, but also choose to stand around on dung, the better to attract the flies and other insects on which these beetles prey. Some South African plants, in the Asclepiadaceae, produce flowers that are not only foul-smelling, but also look very much like the excreta of vultures, doubtless the more effectively to attract the flies they need for their pollination.

Sometimes additional colors may embellish droppings, their pigments deriving from items in the diet as well as from the greenish-brown bile pigments that result from the breakdown of hemoglobin. (Chemists have been able to identify coproporphyrin and bile-acid residues not only in the scats of contemporary creatures, but even in the coprolites of extinct crocodiles and other animals.) Dog and cat owners know that the more red meat their pets eat, the darker the feces, because myoglobin, which gives its color to red muscle tissue, is degraded by digestive systems to brown stercobilin. Humans often produce red feces after drinking much red wine or eating beetroot or Indian pokeberries, which contain an anthocyanin-like pigment remarkably resistant to human digestion. (In certain individuals, the ingestion of beets also colors the urine.) Bears' excreta may be blue if they have been eating blueberries, black if they have been eating flesh, and brown if they've been feeding on acorns or other vegetable matter. The more pinewood a porcupine eats, generally the redder its feces. Hyena droppings, usually green when fresh, tend to go white with age. That is largely due to solar bleaching and desiccation, which expose crushed bone fragments, whereas the whitening of deer scats is more likely attributable to a weft of fungal filaments growing on their surfaces, and that of voles to the dried residues of their urine. Lichens in their diet may confer a bright orange color to the diminutive pellets of pickas. A purple coloration in the excreta of martens or wood pigeons in autumn generally indicates a diet rich in ivy or other berries, while an orange-red hue in penguin or sea-lion feces bears evidence that the animals have been feeding on krill (crustaceans which in turn had taken their carotenoid color from algae in plankton), and the feces of an octopus that has dined on squids tend to be tinged with the imperial purple of their ink. A greenish coloration in penguin droppings suggests that the birds are unhealthy and, as a result of accelerated throughput, are excreting bile pigments. However, the excretions of browsing animals and birds such as geese tend to be greenish because of the persistence of the plant pigment chlorophyll or its degradation products. Grass contains little protein nitrogen, so the droppings of geese, which like herbivorous mammals graze in pastures, are usually dryish, olivaceous cylinders because they contain so much undigested cellulosic material.

When feeding, geese produce a pellet every three or four minutes (flamingos may do so even more frequently). In eighteenth-century France, "merde d'oie" was for some reason a popular color for harpsichords. Birds with less than commendable sanitary habits, including some kinds of hornbill, often stain their eggs with fecal olive or brown blotches that may serve to conceal them from predators. (A remarkable claim has been made recently by Finnish scientists that the dim blue fluorescence of vole droppings, when irradiated by ultraviolet light, may help kestrels to locate the rodents on which they prey, but so far the evidence is somewhat shaky. This reminds me of an imaginative statement by the parasitologist R. S. Desowitz: "if feces were fluorescent [I assume he means "phosphorescent"] just about all of the tropics would glow at night.")

When flushed from their nests, some ducks—eider and tufted, for instance—defecate on their eggs, perhaps thereby rendering them less attractive items for the breakfasts of foxes. (The female eider duck, having incubated continuously for three or four weeks without relieving herself, can produce particularly noisome feces.) We may presume that it was Carolus Linnaeus's experiences in northern Scandinavia, where he must have been subjected to the indignity of being dive-bombed by nesting jaegers, that led him to give this graceful gull-like bird the Latin name *Stercorarius* (literally, "the shitter"). Fieldfares, too, can dive-bomb enemies with remarkable accuracy, especially in defense of their nests. Vultures may have the same expertise. In 1994 they "splashed" legislators visiting the garbage dump in Cartago, Costa Rica, perhaps to register a protest against government plans to close down the facility where they lived. In the same country, fruits of the *Witheringia* tree have been noted to have a strong laxative effect on solitaires and other birds that feed on them. Diarrhea in turtles, however, perhaps because they tend to be more phlegmatic in temperament, is induced less often by fright than by eating unsuitable food such as mud, which they sometimes ingest when more nutritious material is unavailable.

The so-called marine snow seen by SCUBA divers as they flash their lights through seawater is largely the gentle snowfall of plankton droppings drifting down to the sea floor. In the Weddell Sea, up to 80 percent of the falling particles are fecal. Most bottom-dwelling worms would starve without such a supply of manna. This "snow" is not really white. It is olive-green, largely because like horse droppings it contains —along with some still-living plankton cells— pheophytin pigments, the degradation products of chlorophyll. It only looks snowy in the divers' lamp beams. However, some of these pellets have so many luminous bacteria growing on their organic residues that they glow in the dark, and certain midwater animals tend to be attracted to these shining morsels. The constant rain of marine feces may have been more important to the evolution of life on earth than we might suppose. In the Cambrian era, as larger animals evolved, they produced larger fecal pellets, which fell more rapidly through the seawater and thereby accelerated the removal of mineral components such as carbonates and phosphates from the surface layers and their accumulation in the sediments.

Humans excrete 20 to 1,500 grams of feces daily (European values are usually in the range of 100 to 200 grams), depending of course on dietary intake. Healthy human feces are normally about 80 percent water; recorded values range from less than 50 percent (extreme constipation) to over 90 percent (diarrhea). These are not unimportant data: armies (for example) have to make appropriate calculations and arrangements. (For further details the reader is referred to a comprehensive work on the subject by Coughlan et al., 1983.) Of the solid matter (at least that produced in most European countries), most consists of insoluble fibrous materials; within limits, the more fiber in our diets, the better, since it effectively facilitates and tends to regularize elimination. So, paradoxically, even the indigestible components of our food are good for us. A high-fiber diet, like that of many African tribes, may reduce the incidence of colon cancer, possibly by accelerating the passage of fecal matter through the gut and thereby reducing the time available for toxic fermentation products to accumulate. The fibrous fraction is accompanied by fats (in some cases exceeding 10 percent), salts, cells sloughed off from the linings of the intestines, and enormous quantities of bacteria (variously estimated to be 30 to 80 percent of the total bulk). Our usual production depends on how much we have eaten in the preceding forty-eight hours; over an average lifetime, the linear production of an average human being would total several kilometers.

Some carnivorous mammals, birds such as vultures, hawks and owls, and shore birds that feed on shellfish, regurgitate the more insoluble items of their diets, including bones, beaks, fur, and shells, rather than allow them to pass through the gut and complicate digestion and defecation. One may assume that in the end this is a much more comfortable arrangement. If indigestible matter such as animal hair or plant fiber accumulates in the stomach, some of it may become compacted and concreted to form what are called "bezoars"; under other conditions, lumps of insoluble matter formed in the intestines are called "enteroliths." In some cases, such concretions are considered to have medicinal or even magical properties.

The scats of aardwolves are often full of sand, taken in along with the termites on which these animals feed. Those of civet cats may sometimes consist largely of the cylindrical carapaces of millipedes, while those of lions, rather surprisingly, may contain the quills of porcupines. A gold wedding ring once turned up in a bobcat scat, though how it got there is anyone's guess.

Scats exhibit many different forms and textures, as most city pedestrians know, reflecting the varied nature of the solids in the animals' diets. These residues provide valuable information to field naturalists, who need to sort, identify, and evaluate the various elements. Some ecologists in Colorado (Bryant and Williams-Dean, 1975) have developed a simple technique for washing coyote scats before sorting the insoluble components. They tie them in sections of nylon stockings and run them through a washing machine (presumably in the absence of the more usual domestic items of apparel). In such studies, one should bear in mind that some of the components of carnivore feces may have originated from the gut contents of their prey. That's how grass and seeds ingested by zebras, say, eventually find their way into lion droppings.

Horse droppings tend to be full of chewed-up straw and chaff, which confer on them a firm texture. Cattle, on the other hand, being ruminants, do a lot more chewing and rechewing to digest their food, and many bacteria in their rumens have the ability to liquefy cellulose. As a result, cow droppings are usually semiliquid when first expelled. Later, as the droppings lie around in the fields and dry out in the sunshine (as they usually do eventually, even in Britain), their texture becomes firmer. They then may reach a stage that could be mistaken by a drunken yokel for a tam-o'-shanter (at least, so a Scottish folksong tells us) and worn on the head instead of a cap, although this is hardly to be recommended.

The excreta of young mammals, before they are weaned, are normally semiliquid, since there is very little solid matter left over from a diet of milk. As every human mother knows, the color and consistency of their babies' feces, which at birth are called "meconium," change at the time of weaning. The manufacturers of disposable diapers have to study such matters, which are thus of considerable economic importance.

The rheology of adult human feces, too, has occupied the attention of scientists: it's all grist to the mill of knowledge. When faced with the problem of evaluating the flushing efficiency of different models of domestic toilets, the technicians of Consumers Union in the United States felt it necessary to devise a sort of synthetic turd to simulate the natural product, and so concocted cylinders of a mixture of sawdust, flour, shortening, and just enough hollow plastic beads to confer a slight degree of buoyancy. The proportions in this recipe were not published—perhaps they could be varied over wide ranges—and one may assume that no cooking was needed. Mycologists studying the development of dung fungi have dubbed such artificial feces "copromes."

Since swimming marine animals have no need to conserve water, provided they can find some way to eliminate salt, their feces are usually more or less fluid, with a water content exceeding 90 percent. Some seals and whales may normally have diarrhea. However, the excreta of leopard seals at certain seasons are filled with the feathers of luckless penguins, while those of Weddell seals are sufficiently solid to accumulate on the bottom of the ocean off Antarctica, where, *faute de mieux,* they provide a welcome source of nourishment for the local starfish. The feces of parrot fish, which nibble coral around many tropical reefs, are so full of limy material that they contribute in no small measure to the white sands often associated with such locations. A good-sized fish may deposit almost a kilogram per day, often defecating in a hollow a little distance away from its customary feeding area. A half-dozen large fish, nibbling away steadily at a reef, can easily generate a ton of coral sand in a year. (Sea bass, like us, defecate in a diurnal rhythm, usually shortly after breakfast.) When aquarium fish trail long streamers of fecal matter from their cloacae, this is probably a consequence of a constipating diet.

Unlike the relatively bulky feces of earthworms, those of snails are much more compact, more cohesive, and less friable. This is because their diet, consisting of leaves and other vegetable matter, is more nutritious, and because their digestive systems produce special enzymes for breaking down the molecules of cellulose. Snails are almost unique in this respect: few other animals can digest cellulose without the help of intestinal microbes.

The first act of a butterfly, even before its wings have dried sufficiently for flying, is to excrete a droplet of liquid feces (often reddish in color) to relieve itself of the fluid remains of its earlier incarnations as caterpillar and pupa. Like the fecal matter of newborn mammals, it is called meconium. Newly hatched mosquitoes, which still retain meconium, can be thereby distinguished from older insects that have shed it, a distinction of considerable importance to entomologists involved in malaria research.

4

COPROLITES

Anthropologists insist that only samples taken from the intestines of mummified human remains, or feces containing evidence for exclusively human parasites, can be assigned certain human status. Their further classification of coprolites is made on the basis of form, color, and visible contents (Fry, 1977). For instance, the presence of teeth from which the enamel has dissolved suggests that the feces may have been those of crocodiles, whose digestive juices are known to be particularly acidic and whose digestive transit times tend to be relatively slow. Although few paleontologists collect them even now, as early as 1823 Dean W. Buckland identified certain coprolites, from caves near Lyme Regis on the Dorset coast, as fecal products of hyenas, and many hundreds of scientific articles have been published since then. Coprolites have been reported from every continent except Antarctica, and from every epoch since the Paleozoic. There are large collections in scientific museums in California and New Mexico, but relatively few elsewhere. (For a good recent review of the subject, see Hunt, Chin, and Lockley, 1994.)

An illuminating account of recent studies of prehistoric human excrements was published a few years ago by Bryant and Williams-Dean (1975). They reported that, when soaked for three days in a strong solution of trisodium phosphate (more usually employed as a paint remover), coprolites of omnivores such as man and the coatimundi render cloudy extracts, darkly colored like strong coffee, whereas those of herbivores and carnivores yield clear brews, their color more like weak tea. Items sorted from the solid residues—bits of shell and bone, fish scales, chips of charcoal, seeds, pollen grains (of special diagnostic value), and fibers of various sorts—have provided valuable clues to the diets of the animals that excreted them. Much of the research on human coprolites has been done with Mexican material, preserved by drought. However, data from a fine collection of some 500 samples from the south of France have proved to be among the most fruitful, as have 146 from dry caves and shelters in Utah.

In some parts of a cave near Lovelock, Nevada, the deposits are piled to depths of more than 7 centimeters, indicating that those sections had long been used as latrines. Among other insoluble items, these coprolites have proved to contain not only seeds and other plant debris but also beetle wings, indicating that the hungry cave occupants, like their ape ancestors, were sometimes hungry enough to include even large insects in their sparse diets (Fry, 1977). A remarkable coprolite 8 to 10 centimeters wide, presumed to have come from an eighteenth-century Chumash Indian fisherman, was disinterred from a marine-sediment sample collected a few years ago in the Central California Basin. It contained bits of fishbones, perhaps chewed by the fisherman (who, one assumes, had subsequently relieved himself over the side of his boat), and had evidently retained its integrity in the anaerobic mud long enough to become hardened by being impregnated with apatite. From such fossils we have learned a lot about the diets of our ancestors, as well as about their internal parasites whose eggs are commonly found in coprolites. Some of the components of ancient human coprolites from La Quinta, near Palm Springs, California, are sufficiently well preserved to have yielded small samples of DNA, which, by modern techniques, have provided indications of the sex of the fecant.

Some pack-rat droppings, which accumulate for centuries in dry caves, are indurated by a sort of amber that forms when the animals' urine evaporates, thus helping to preserve identifiable components of their diets. The accumulated feces of *Megatherium,* not really fossilized but preserved by desiccation in certain dry American caves (some only a few miles from Las Vegas, Nevada), tell us a great deal about the diets of these enormous giant sloths, now alas extinct. Residues of more than seventy genera of plants have been identified in their dung. Using state-of-the-art molecular biology (PCR sequencing), Poinar and his colleagues (1998) have been able to identify the DNA of hillside plants in many families, indicating that those sloths lived on leafy salads condimented with mustard and capers. In Arizona caves, similar deposits of mammoth dung, more than thirteen thousand years old, contain evidence that these huge animals may have had to feed on tundra plants much smaller than the trees on which contemporary elephants browse. Perhaps the most interesting coprolites of all, the fossilized droppings of dinosaurs, are also a wonderful source of information about the diets of those long-extinct reptiles. Some of the biggest dinosaur coprolites exceed 12 centimeters in length: a few are flattened, compressed (certainly after excretion) to the thickness of cardboard. So far, the record for size is held by a recently excavated coprolite of a tyrannosaur from Saskatchewan, which had evidently chewed up a smaller dinosaur before defecation. Despite a little fraying at the sides, after being dug up it is still some 44 centimeters long and 16 centimeters wide, and it weighs over 7 kilograms. Many coprolites of extinct vegetarian reptiles contain fragments of fern and cycad leaves, but naturally no grasses or skins of fruits, etc., because higher plants had not yet appeared on earth. They are not only scientifically interesting, they also have monetary value: in a 1993 London auction, twenty-three pieces of dinosaur dung from Hanksville, Utah, were sold for $4,500. Some examples are illustrated in an article on dinosaurs published in *National Geographic* magazine (183, 39; January 1993).

In parts of Bedfordshire, near Shillington, such large accumulations of coprolites were found that they were still being mined early in this century as a source of phosphatic fertilizer. In the coprolites of mastodons and smaller animals that lived eighty million years ago one can even find fossilized bacteria. And one of the most scientifically valuable coprolites was discovered recently in Inner Mongolia: when cracked open, it proved to contain teeth of the most primitive known rodent—eaten, some fifty million years ago, by who knows what?

Another surprising discovery was made by divers exploring underground caves in Grand Bahama: they found patches of fossilized chiropterite (bat droppings). Obviously these had not been produced by submarine bats. They were in fact laid down by aerial bats, flitting among the stalagmites in limestone caves that were only later submerged by a rise in the water level of the Atlantic Ocean.

Some creationists claim that they have found coprolites of unidentified animals in a deposit that others have estimated to be more than one hundred million years old, at the site of Noah's ark, in Ağri, eastern Turkey. Of course, evidence for the ark is less solid than for the coprolites.

What might be called microcoprolites—the pellets of planktonic crustaceans, etc.—have been exhaustively studied in deposits brought up from the bottom of Lake Tanganyika. The sludge contains more than a million per gram. They are full of the shells of various species of diatoms, which provide evidence for climatic conditions prevailing in the past fourteen thousand years. And among the earliest known coprolites are tiny pellets, barely one thousandth of an inch long, perhaps produced by some of the first terrestrial arthropods (mites or creatures of similar size), found in Upper Silurian deposits of sandstone. Pellets found in even earlier sediments, dating from about eight hundred million years ago, have been suspected of being coprolites produced by some of the earliest invertebrates on earth.

5

SMELLS AND OTHER CHEMICAL COMPONENTS, INCLUDING GASES

Fecal odors vary with species and diet. Understandably, those of otters are especially fishy. The smells also vary with age and state of health. Mostly they are regarded as unpleasant; indeed, we are told that the scientists who worked out the chemical composition of skatole, a predominant odoriferous component, had to carry out some of their studies well away from laboratories occupied by colleagues, and to burn their lab coats at the end of their investigations. (A "Fecal Odor Eliminator," now marketed in handy plastic bottles by a company in New York, was presumably not available at the time.) Other odorous substances include the related compound indole and various sulfur-containing molecules like H_2S and mercaptans. Notorious in this connection are the smells of garlic and onions (allyl and vinyl sulfides), which seem to pass unchanged through the human digestive system, whereas others, especially from such foods as beans and crucifers (e.g., cabbage), are generated by anaerobic bacteria in the gut (Moore, 1985; Sastry et al., 1980). Many of these smelly substances may later turn up in analyses of water from sewage effluents downstream. Changes of fecal odors constitute the subject of frequent discussions among mothers of young babies, as they were reputedly among the medical advisers of Chinese emperors. Rabelais related how some young ladies—they may have been novitiates in the Thélème Abbey, but I have not been able to confirm this—had such pure natures that their little virginal feces were small and fragrant. The phenomenon may be questioned, since it is hard to suggest a physiological basis for such a manifestation of piety. It may sound like pure fiction, but in fact scientists have recently discovered that the excretions of young virgin queen bees (aged between one and fourteen days) are actually fragrant, whereas those of other bees are not. The perfume of such a virgin queen in some way controls the behavior of her subjects in the hive.

Females of the black-footed salamander are said to select their mates at least partly by the texture and smell of their feces, which reveal the kinds of food that a nubile male is able to catch and eat, as indications of his levels of energy and predatory efficiency.

Young horses can recognize their mother's feces by smell, and stallions can distinguish the dominance hierarchy of other male horses in the same way. Similar phenomena have been documented among wild rabbits and house mice

(Goodrich et al., 1981; Leon, 1980): some of these fecal odors can even influence their heartbeat rates. Many adult mammals, especially carnivores such as mongooses and aardwolves, have special anal glands whose secretions add to the characteristic odor of their feces, which presumably plays an important social role, like that of urine, in the marking of their territories (see Decker, Ringelberg, and White, 1992, and Chapter 6). It is reported that wolverines, when they have eaten as much as they can stomach of a dead carcass, sometimes mark the remains for later consumption by defecating on it, which confirms the general impression that these beasts are not particularly fastidious eaters. Droppings of hyraxes are further scented by the urine that these animals often excrete over them. Smells serve also in pack recognition, which is perhaps why wildebeest and young hunting dogs are prone to roll in their own excretions, the better to indicate to which family they belong. (Priam, king of Troy, rolled in a dung heap on learning of the death of his son Hector, perhaps because he felt that grief can be expressed better by odorous materials than by just sack cloth and ashes.) Some elephant hunters have tried smearing themselves with elephant dung as an olfactory camouflage; its efficacy is questionable. Lions have been observed to do the same thing—but why? They surely are not planning to hunt elephants. And why some dogs choose to roll in other smelly materials, not just fecal matter, is also not immediately obvious. Their olfactory preferences must be very different from ours. However, there are possibilities for perfuming dog droppings *in vivo* so that they become less offensive, at least to us. Special dietary additives for this purpose are already on the market in Beijing (where private ownership of dogs has to be clandestine, because it is formally illegal).

The scats of pine martens smell musky, not unpleasant to some, and quite unlike the more fetid excreta of their relatives the beech martens, the difference perhaps being due to the aromatic oils of the leaves among which they live. The brown droppings of rabbits may have a relatively sweet smell, and the scats of many animals such as deer have characteristic odors in the rutting season, distinguishable by experienced scatologists as well as by the animals themselves (Halfpenny and Biesiot, 1986). Voles, understandably, recognize and tend to avoid the droppings of foxes that normally prey on them, evidently preferring not to become components of the next meal. Similarly, cattle and sheep tend to avoid food tainted by the smell of coyote or panther droppings, and monkeys shun that smelling of wild cats and panthers. (A project to market lion droppings in the Netherlands, to be distributed as a deterrent to the invasion of gardens by neighborhood cats, proved unsuccessful, perhaps because Dutch cats are as a rule unaware of the size and ferocity of their feline cousins in Africa.) In March 1996, symposium workshops were held under university auspices in Wetunka and Woodward, Oklahoma, specifically to discuss pig manure and its characteristic odors. Recently, English farmers have been deliberately scheduling the dunging of their pastures shortly before the announced arrival of hordes of vagrant "travelers" (gypsies, etc.), usually with a considerable deterrent effect. Following this example, perhaps, a farmer in Woodstock, New York, successfully repelled a flock of hippies, anxious to celebrate a twentieth reunion on his land, by the judicious and timely application of a few tons of chicken manure.

Cows generally prefer not to graze in the immediate vicinity of fresh cowpats for several weeks, presumably (and understandably!) repelled by the smell. During this period of avoidance, according to Japanese researchers, the growth of clover in such areas is promoted, which further enhances the fertilizing effect of the droppings. Experiments with fluorescent dyes have shown that some components of the cowpats' solid matter as well as their smells are dispersed for at least a couple of yards around the pats, and remain repellent for at least a couple of weeks. However, cowpat smells are attractive to certain flies, which are stimulated to lay eggs on them. Experiments have shown that anosmic flies, surgically deprived of their antennal organs of smell, lay fewer eggs and take longer to do so. A lady in Brazil who made a special study of insects that are attracted to human feces noted that after each change of diet our excreta may appeal to different species of flies.

The Greek word *kopros*, which has the same meaning as scat, appears in the name of certain shrubs in the coffee family: *Coprosma* flowers have a characteristic fecal odor, attractive to pollinating flies. Similar odors are produced, doubtless also to attract pollinating insects, by flower spikes of certain plants in the family

Araceae (to which the skunk cabbages and the "voodoo lily" belong) as well as by fungi like the stinkhorns and even by a coprophilous moss, *Splachnum*. Some of these smelly components have been found to be toxic, at least to rats. Their presence has also served to confirm that a part of an ancient Roman archaeological site in Stanwick, Northumberland, was in fact used as a cesspit.

For diagnostic purposes, as well as for tracking, it is sometimes useful to determine the pH of a fecal sample, or its content of protein or of various bile acids. Certain bird droppings have special chemical features. Whereas it is almost impossible for us humans to distinguish male from female parrots, one way in which this can be done is by analyzing the steroid hormones in their excretions. For some well-equipped analytical laboratories, parrot sexing may prove quite a profitable business. Also, if one really wishes to do so, one can detect pregnancy in a rhinoceros by the presence of progesterone in her droppings. The steroid coprostanol (formed from cholesterol by intestinal anaerobic bacteria) is so stable that it is used as an indication of pollution by predominantly human feces when waters are being evaluated for domestic use.

In a recent study of the droppings of bears in the mountains of Italy, where they are now almost extinct, it was found that individual animals could be readily identified by the molecular biological signatures of cells from their intestinal systems. Similar studies are now being carried out on African baboons and elephants because this is a less hazardous undertaking than scraping off bits of their skin *in vivo*. And when locusts change from the solitary to the gregarious phase (the "hoppers" that are sometimes responsible for devastating plagues), a corresponding change can be detected in the hormones in their excreta.

I once saw mounds of vampire-bat feces in underground caves in Costa Rica. Unlike insect-eating bats, whose droppings are packed with the undigested legs and wings of moths and beetles they have eaten, vampires live solely on a diet of blood, which consists mainly of water, salts, and hemoglobin. The globin part of the latter molecule is digested, and the hemo part, converted largely to stercobilin, is excreted as a sort of black tar that accumulates in huge heaps under vampire roosts. Evidently few microbes can take advantage of this accumulating dung pile, which just keeps accreting, perhaps for hundreds of years. And consider the implications: every kilogram of this hemolike material must have originated from about a ton of blood. Birds called swiftlets, which occupy different sections in some of the same caves, produce very different kinds of droppings, obviously reflecting their wholly dissimilar diets.

The droppings of adults and caterpillars of the monarch butterfly, which feed on milkweed leaves, contain toxic cardenolides derived from the sap of these plants, and so do the insects themselves, which many birds consequently find so distasteful that they avoid eating them. The renowned nineteenth-century French entomologist Jean Henri Fabre wondered whether the irritant hairs of many colonial caterpillars might derive their urticant nature from the insects' droppings, which somehow get rubbed on their furry coats in the course of their movements. Perhaps the chemicals originate from toxins in the leaves on which the caterpillars feed.

In addition to solid matter, most excretions include gases, technically known as flatus and, more vulgarly, as farts. (Eructations, the burp and belch gases produced in the upper part of the digestive system, fall outside the scope of this book.) "Excessive gas is considered by many to be one of the most common afflictions of mankind" (Levitt, Bond, and Levitt, 1981), though I think "affliction" is perhaps too strong a word to use for so common a phenomenon. One of the major gaseous components of flatus, not only of ruminants like cows and yaks and of such vegetarian reptiles as tortoises but also of ourselves, is marsh gas, or methane, which, along with larger amounts of another combustible gas, hydrogen, is formed by anaerobic bacteria. In horses this takes place mainly in the lower regions of the gut, while in ruminants like cows most of it is "burped" out at the front end. Both of these gases are odorless when pure (which they rarely are in human flatus). Our guts normally contain around 100 milliliters of this gas mixture, and our daily rates of expulsion may reach 2 or 3 liters. The sound of intestinal gas bubbles, burbling their way down and out—in apparent defiance of the laws of gravity—is called "borborygmus" (a fine example of onomatopoeia). I suggest it might also be known as colophonics.

Methane production is particularly high in

those of us who had two "productive" parents, and it is increased by the consumption of undigestible materials, such as oligosaccharides, which reach the lower bowel and ferment there. Baked beans and certain other legumes are well known to raise the methane content of normal human flatus from only a few percent to almost 25 percent. In recent years, there has been a lot of scientific dietary research toward reducing this effect, reportedly with some partial success. Some human excrements, from people who have eaten a lot of cellulosic material (especially beans) or who are infected with internal parasites like *Giardia,* incorporate so much gas that, like beaver droppings, they float, which makes for complications in operating the settling tanks of sewage plants. In the vicinity of coastal towns with inadequate disposal systems, this feature may understandably discourage people from swimming in the ocean.

It has been estimated that a cow may produce several hundred liters of methane per day, and that sixty million tons of this gas, or about 15 percent of the total annual production entering the world's atmosphere, comes from cattle. The effects of all this methane on the so-called greenhouse effect—the warming of the earth's atmosphere—are not inconsiderable, and it has been reported that Australia may consider taxing the gaseous effusions from cattle, though how their volumes are to be monitored is far from clear. Another 5 percent of the total methane production, rather surprisingly, originates in the guts of termites and locusts. How much is formed in the intestines of marine zooplankton is now a hot subject in oceanography. As one of the harbingers that "summer is y-cumen in," thirteenth-century farmers noted, and sang, that the "bucke ferteth" (i.e., the buck farts), presumably a consequence of the renewed abundance of fresh green browse.

The Chinese have a saying, *Xiang pi bu tsou, tsou pi bu xiang*—anal emissions may be loud but odorless, or silent but smelly. (Some Chinese sayings have political implications; if this has any, they escape me.) There are undoubtedly countless references in the literature and on the stage, and in schools, pubs and bars, bedrooms, and living rooms everywhere, to these sound effects, but they are mostly ribald. Suffice it to mention here that higher tones are associated with greater pressure differentials, while both duration and tonal frequency increase as the size of the orifice is reduced. The farts of a mythical beast, the bonnacon (which may have resembled a large goat), were reputed to smell so foul that they not only repelled pursuing predators but even blasted vegetation, leading one to wonder which dietary components could have been responsible for this effect. In one medieval illustration of this feature, the emissions were colored blue. And, on the subject of mythology, we should also mention a story, which doubtless amused Indians of the Winnebago tribe, about the Great Magician blowing up the whole of mankind with a single fart, and then covering the surface of the earth with the feces that followed it.

In crowded and poorly ventilated cowsheds, kept closed overnight, accumulations of methane can sometimes form explosive mixtures with the oxygen in the air, and cattlemen have to be careful when and where they light matches. There have also been reports of explosions caused by sparks in hospital operating rooms where electric cautery was used in operations on a gas-filled colon. Many schoolboys—possibly also schoolgirls; I don't know about this—have much fun exploiting the combustibility of farts, especially when demonstrated in a darkened room for the *son et lumière* effect. Doubtless many of our ancestors learned how to make this magic back in their early cave days. However, assertions that such gases are sometimes capable of spontaneous combustion, like *ignis fatuus,* are of course wholly without foundation.

The use of sewage methane as a fuel is reviewed in Chapter 16.

6

TERRITORIAL MARKERS AND TOILET AREAS

Brown rats, unlike their black cousins, generally leave droppings in the corners of rooms and barns, or along the walls of buildings. Water voles mark their riverside runs in much the same way. On the Peruvian altiplano, and at zoos and farms elsewhere, llamas seem to designate special toilet areas, where thousands of little turds are dropped on limited sites only a meter or two across. (Hyraxes and rabbits tend to do likewise, but the more solitary hares don't seem to bother.) Indeed, llamas sometimes wait patiently to take their turns at the latrine areas. I doubt whether they do this for purely hygienic ends, since their habitat is so wide and windy. (I am told that domestic pigs can be easily housebroken—or sty-broken, perhaps—if their facilities are adequately designed and plumbed. This is probably old news to today's swineherds.) Llamas, vicuñas, and their kind use droppings to mark the territories of their flocks, if in fact they have ranges that are restricted in this way. Alternatively, they may behave like wildebeest on the plains of Africa, which set off certain patches away from their browsing areas as latrines. In this way, they effect a sort of rotation of crops, as first one patch and then another is fertilized, eventually to produce better grazing. A team of scientists in a statewide survey of grazing animals in the New South Wales outback is currently using satellites and a global positioning system to localize and map the distribution of clumps of kangaroo, sheep, and goat droppings. (A NASA satellite called "Quik-Scat" monitors ocean wind; it doesn't monitor scats as we know them.)

Under some conditions we humans, too, tend to confine our excretory activities to limited areas. A study in West Bengal indicated that villagers prefer to confine their outdoor defecation to certain socially recognized sites, representing only a small percentage of the total area around each village. The operation is usually accomplished relatively quickly, generally by men in the mornings and by women in the afternoons.

Many kinds of antelope mark their ranges with dung middens. Hippos indicate their territorial limits by dispersing their sloshy feces with rapid to-and-fro wagging of their tails. (It is a poorly kept secret that many trains, as they speed between cities, dispose of their human waste by spraying it around in much the same way as do hippos, though doubtless not for the same reason. I am informed that for many years now Amtrak trains in America have installed a more hygienic system, and one could hope that this might be generally true elsewhere.) Male black buck often sit on their feces, the better to establish their territorial rights.

Gorillas, which in nature tend to foul their arboreal nests, presumably do not have to worry much about domestic sanitation because they usually move on to another resting place every night. But sloths, which defecate only about once a week, slowly climb down to the foot of their tree to do so. (They are leisurely not only in their gait but also in their metabolism. Food may take as long as six weeks to pass through the digestive system of the two-toed species. Compare this with the twenty-minute passage time recorded for several species of bat.) Some sloths build a pungent midden, which might serve to indicate a "trysting place," not only for the sloths themselves but also for moths that infest their fur. The three-toed sloth often goes to the trouble of burying his (or her) droppings in a shallow depression dug with his stumpy tail, and then carefully covers them with soil and leaves, taking about half an hour to finish the job properly. Goodness knows why. One is reminded of Saint Simeon, who chose to express his Christian piety in ascetic solitude atop a sixty-foot column, and who at weekly intervals lowered his excrements to the ground by the bucketful.

Sometimes a marten, on finding a pile of scats left by another marten, may add a personal contribution to the heap, though whether as a challenge to the territorial claimant or as a kind of social visiting card is a matter of speculation. Bears have been reported sometimes to defecate on top of cowpats, and cats or dogs on anthills, though they could hardly be disputing territories in this way. The more social spotted hyenas have been seen defecating collectively during scent-marking patrols around their territorial boundaries. Badgers generally scrape shallow toilets in the ground, presumably for the same purpose, although underground, inside their setts, they set aside special side chambers as toilets. They have been found to create two kinds of "latrines," larger ones for feces with special anal smells (saying, presumably, "keep off"), and smaller ones for casual excretions. Likewise, African mole rats, living in enormously long, branched burrows, designate certain side tunnels specifically as toilets which, one may suppose, can be walled off when they become too noisome. Caterpillars of the *Strathmopoda* moth, which live in tunnels in the mud of mangrove swamps, have been observed to do likewise.

Marmots, muskrats, foxes, otters, civets, hyenas, and badgers mark their home ranges with turds (otter droppings are known in the trade as "spraints"). Ecologists in England have been able to determine the home ranges of badgers by putting out for them peanut-butter bait containing variously colored bits of indigestible plastic, which can later be identified as individual markers in their droppings. Wombats in Australia also mark their home ranges. I recall seeing little heaps of droppings on every raised mound or log along a nature walk (and wombat run) near Sydney. However, wombats must produce more droppings than they need for marking their territories, for in some of the larger and older cave-like sections of their communal warrens, their fecal accumulations are knee-deep. (But note: wombats' knees are much lower than ours.) Large Australian lizards called goannas are reported to mark their territories in the same way, as do even lowlier creatures such as ants.

Some middens marking territorial limits can be quite impressive: those of rhinoceros have been recorded up to 5 meters in diameter. The rhino scatters them in an apparently deliberate fashion, and before depositing a fresh load of dung, a male will often disperse the older deposits with his horn, perhaps further to enhance the effluvium. (As Addison wrote in *An Essay on Virgil's Georgics*—though not about a rhinoceros—"He breaks the clods and tosses the dung about with an air of gracefulness.") Such middens are often more evident than their producers: a zoology student at Oxford who spent several years ostensibly studying the ecology of rhinoceros in Africa never actually saw any of the animals themselves, only their traces.

The feces of baby songbirds fed on diets rich in insects are often characterized not only by color (usually white) but also by size, shape, and texture, being packaged into soft-skinned but often tough mucus-coated lozenges that can be easily removed by the parents. These lozenges are formed only by the younger nestlings: the excreta of older bluetit nestlings, for instance, lack the characteristic capsule. The pellets are generally released just after feeding, their production—one or two every hour—being sometimes promoted by parental prodding. Juvenile pellets of the diminutive bushtits are proportionately among the largest known. Many avian parents take them directly as they emerge from the cloacae of their offspring, eating those first produced and carrying away the later, larger products. Others merely pick them up and dispose of them shortly after they are excreted. The dipper, and many species of swallow, carry away such pellets and drop them in a nearby stream or pond, if one is available; lyrebirds may do the same, or even bury them in specially excavated holes. Parent birds fostering young cuckoos perform the same services for their adopted offspring. Older juvenile birds often take it upon themselves to drop their excretions over the edge of the nest; young gulls sometimes walk a few paces away from the nest site before defecating. Baby hornbills and kingfishers have a neat trick of shooting their pellets out of the nest hole in which they are raised, aiming at the light with surprising accuracy. One way or another, most birds manage to avoid fouling their own nests.

Some caterpillars flip away their feces with a characteristic movement of their back ends. (The ballistics of flung feces is a special subject. The wife of the Scottish poet Robert Burns was reputed to have had a special knack of throwing cow dung against the walls of buildings.) Some apes in zoos throw their turds at spectators, probably to elicit shrieks from the human visitors, but almost certainly this is a learned trick and not a natural mode of defense. Hawks and other raptors, and some kinds of penguins, squirt their feces with considerable velocity, raising their tails to keep their plumage from becoming sullied. On the other hand (or foot?), in hot weather African wood storks and some New World vultures deliberately foul their legs in apparent attempts to cool them. Defecation may also serve to lighten the aerodynamic burden in flight: snipe often precede their spectacular upward nuptial flights by first shedding some excretory ballast. But why loons often bother to clamber up to a lakeside bank before defecating is a bit of a mystery. Perhaps they do so to avoid sullying the water in which they have to seek fish.

7

DEFECATION ATTITUDES, HUMAN AND ANIMAL

Humans tend to adopt a characteristic posture for defecation. Some people sit, others squat. In the countryside, it may be expedient to change positions or attitudes with the seasons. A wintry period in December and January, when the ground in western Canada is colder than usual, was (perhaps still is) designated by the Haida Indians as Kong Kyaangaas, or "stand-up-to-shit month." A Hindu codex prescribes that one face north in the mornings and south in the evenings, thereby ensuring that one's shadow falls to the left; I can't think why.

Many terrestrial mammals stop and lower their hindquarters somewhat on defecation; baboons, small carnivores, and cavies generally squat. Sometimes this permits the droppings to stand on end, since they have not far to fall. (Incidentally, baboon droppings seem to be a favorite object of aerial play among eagles of the Serengeti Plain.)

Bigger animals, like cows and horses, might find that squatting entails too much effort, and the time that they would need for getting up could delay their running away when danger threatens, so they generally defecate while standing or, like horses, without even breaking their stride. Unlike most dogs, cats often scrape soil over their feces to bury them; neurologists have discovered that the whole business of digging, squatting, defecating, and scraping is a reflex sequence that can be stimulated electrically even in an animal that is technically dead. In the wild, cats are said to do so only within their hunting territories, whereas domestic cats in many homes are provided for this purpose with cat litter (at today's prices, costing about 10 cents per pound). While excreting, cats and dogs seem to wear an expression indicating that nothing could be further from their minds.

Flying birds generally seem to have no problem in releasing their droppings while in flight; however, I don't know about running birds like ostriches. Most kinds of bats, since they hang upside down, adopt special positions when excreting in order to avoid fouling their fur. In the trees where they roost, fruit bats generally arrange themselves in a sort of hierarchy in which the dominant animals occupy the highest branches, thereby presumably obviating or at least reducing the frequency of fecal pellets falling on them.

8

HUMAN TOILETS AND CHAMBER POTS

Many Native Americans, including Caribs and Arawaks, like cats and sloths, made a point of burying their feces, while in Asia Minor the Essenes are supposed to have carried a mattock for this specific purpose. Similarly, earlier in this century soldiers engaged in desert warfare used to talk about "going for a walk with a spade" before disappearing behind a convenient dune. Understandably, such operations should be carried out at some distance from areas of human circulation. In his play *Plutus,* Aristophanes has an old priest complain about people coming to his temple to excrete rather than to pray, while in the Domesday Book there is reference to a statutory fine for anyone caught defecating in Chester Cathedral. But though sacrilegious practices of this sort continued in France (in no less elegant palaces than Fontainebleau and Versailles, at least to the seventeenth and eighteenth centuries), one may assume that public disapproval of such actions is fairly general in all societies.

In really cold climates, the disposal of human feces presents no serious problem. In ever-frozen parts of Greenland, they were (and perhaps still are) piled outside houses like so much firewood. But in most inhabited parts of the world, this is not an aesthetic option. Before flush toilets we had cesspits. In medieval England, those entrusted to the unenviable task of periodically cleaning out the accumulated ordure (then called "gong") could earn a pretty penny for their labors. In A.D. 1450, one such person was paid seventeen shillings for removing six tons of the stuff. (If the cesspits were not emptied periodically, but were merely covered over, they could present special problems to those who in later years might decide to build on the site.) Not infrequently, such barrowloads of ordure included matters other than feces. In "The Nun's Priest's Tale" (*Canterbury Tales*), Chaucer recounts how the corpse of a murdered man, concealed under a pile of feces, was smuggled to the city gates in a dung cart.

This reminds me of events on Cape Cod, Massachusetts, where I lived in the early fifties. At several of the weekly town meetings, held in Falmouth in the winter months for want of livelier entertainment, there were heated discussions on the advisability of replacing individual cesspits (such as ours at Woods Hole) with a costly public sewage system. And in the same period, certain individuals who went from door to door with a tank truck, purporting to empty the accumulated fecal matter from our pits (for a fee, of course), were found in fact to do little but insert a pipe and for a few minutes generate an impressive but ineffectual roaring noise while our cess remained *in situ*. In due course, they were caught and prosecuted.

A prototype toilet seat, constructed to stand over a dish of sand, has been found in excavations of an Egyptian house built about 2700 B.C. (A wooden easing stool was also found from that period: the hole, barely six inches across, was too narrow for it to have been designed for birthing.) Around 1350 B.C., when the pharaoh Akhenaton was pushing monotheism, toilet seats were made with carefully curved depressions, indicating that the ancient Egyptians—at least the upper classes—sat on them rather than squatting. The hole was sometimes keyhole-shaped, and where there were no flushing arrangements, the seat was supported over a chamber pot. Royal graves from the Second Dynasty had special toilet chambers, presumably for the convenience of the departed spirits of pharaohs and their entourages.

The portable throne on which, for several hundred years, the Pope was carried in procession at St. John Lateran was called "the stercoraceous chair," presumably because of speculation about what might sometimes be going on under the papal robes. (I have found no details of its design that might refute or support this hypothesis.) Evidently, cleanliness and godliness often go hand in hand: in old Jerusalem (see 1 Kings 14:10) they may have passed together through the Dung Gate. John Calvin has been credited with promoting improvements in the plumbing systems of Geneva; Martin Luther, who is reputed to have suffered from chronic constipation, composed many of his treatises while sitting on the toilet.

For Francis I at Amboise Castle, some five centuries ago, Leonardo da Vinci devised not only water closets, with flushing channels inside the walls, but also folding toilet seats not unlike those in use today. The flush toilet, or jakes, was reputedly devised by Sir John Harington, a godson of the Queen, in the court of Elizabeth the First. In his 1596 book *The Metamorphosis of Ajax* (doubtless a pun on "a jakes"), he recounted how King Richard III planned the murder of his nephews while sitting on a jakes, and how Pope Leo V had been killed in a similar situation. Likewise in *The Arabian Nights* there are references to people being captured or murdered while defecating, though not specifically while sitting on toilets.

In the sixteenth century, when at least one of Sir John's flush toilets was installed at the Queen's palace at Richmond, the royals of England were privileged to sit on closestools with velvet padded seats, some elegantly garnished with ribbons and gilt. King Charles I used a silver potty. In the courts of Louis XIV at Versailles, some were covered with red leather or velvet, while those of Louis XV may have had surfaces of "black lacquer with Japanese landscapes and birds in gold and coloured relief, with inlaid borders of mother-of-pearl, Chinese bronze fittings, red lacquer interior, and a padded seat in green velour." (For many such details I am indebted to Wright [1960], who compiled a generously illustrated history of the bathroom.) I have no information about such arrangements in the royal palaces these days, but I know that some years ago Queen Elizabeth (the Second of Great Britain, but the First of Canada) had access to a two-holer in the suite assigned to her at the faculty club in the University of British Columbia. A similar double facility had been used by one of her forebears, the Earl of Leicester, some 250 years before.

A decent latrine system is of course a sine qua non of civilized society, important not only for aesthetic reasons but also for maintenance of public health. A simple, odorless water-seal latrine, devised under the socialist system of Mao Tse-tung and combined with strict legal injunctions against what Desowitz calls "indiscriminate defecation," has been very successful in reducing the incidence of hookworm infestation, which used to be widespread in China (and still is, alas, all too common throughout the tropics, especially in rice-growing countries).

There are two main types of toilets in use today, the sit-on kind familiar in Western countries, and the squat-on kind, which some racist once considered to be "only suitable for native races." In the flat of my old aunt, in the Latin Quarter of Paris, toilets were (perhaps still are) squatters, whereas my contemporary French cousins have sitters. Incidentally, my aunt also had a bidet, which (according to Larousse) is so called because it is straddled like a pony (also called in French a *bidet*). In Japan and Korea, older buildings that I have visited have only squatter-type toilets, while in many new ones one has choice of either design, sometimes with a little instruction to help the uninitiated. To accommodate diverse preferences for sitting or squatting, many modern buildings in Japan have four kinds of toilet stalls, presenting choices for posture as well as gender.

Harris and Chapman (1984) compiled an amusing little book on the subject, illustrated with an abundance of rural anecdotes and pictures, including a photograph of "the finest lavatory in England, at Hampton Manor, Kidlington." These ladies did not confine their attentions to Cotswold privies, but included many historical details about "lavender men" and toilet arrangements in England during the last five centuries.

The first patent for a mechanically flushing toilet was taken out in 1775 by a watchmaker, Alexander Cummings, but it was only toward the end of the reign of Queen Victoria, more than a century later, that a gentleman fortuitously called Thomas Crapper installed suitably regal toilet facilities at Sandringham Castle. His factory manufactured porcelain toilet bowls with valve-controlled cisterns, some of which are still in use today. Early models may have been less than perfect: Crapper once suffered from a concussion when a mixture of air and sewer gas exploded. (Some fifty years earlier, a plumber, illuminating by candlelight the tenebrous sewage drains under the flagstones of a London house, had suffered a similar fate.) However, Mr. Crapper did not invent the the flush toilet.

In some countries of Europe, many such fixtures are designed to expose solid excreta on a porcelain platform before they are flushed away. This device may originally have had something to do with divination, like the reading of tea leaves, or to facilitate the collection of samples, but more probably it stems simply from a need to conserve water. Most toilet bowls in London, where I grew up, do not have such a shelf, and our feces fell directly into the water, as they still do. I well remember my embarrassment on being told by a ten-year-old classmate about his experiments to reduce the sound of the "plops" so that they would not be heard beyond the toilet door. After flushing, adherent residues, unless they have been allowed to dry, can be readily detached with a handy loo-brush, which is generally disposed nearby in its own stand.

A variety of historic toilets on permanent exhibition in Germany, at the Zentrum für Aussergewöhnliche Museum, in the Bavarian town of Kreuth, near Gmund, includes one of the first flush toilets, an elegant model used by the Crown Prince Rudolf. In Japan, where a third of the households now have flush toilets, the boom in plumbing arrangements has inspired a variety of new styles, some costing as much as $3,000. Many of them are produced by the Toto company, which controls some 60 percent of the national market. In such a country with chilly winters, electrically warmed toilet seats provide a welcome comfort.

Among recent innovations in public toilets is a disposable plastic covering that can be renewed automatically by pressing a button, and a device for the irradiation of the toilet seat with ultraviolet light which, one hopes, causes more death than mutation among whatever bacteria may occur there. This subject (what the French call *selles,* "saddles") is too vast to be discussed further here. However, I cannot refrain from quoting an English art museum curator, guiding a French visitor around an exhibition, who referred to an equestrian portrait of King George in the saddle as *à la selle.*

More than one patent has been taken out for a toilet designed especially for receiving and disposing of feces in the zero-gravity conditions of space travel. In such conditions, canister collection and compaction are more acceptable to crews than other suggested systems, and "the waste-collection equipment should retain waste and paper for at least 210 person-days, and should be quickly and easily replaceable in flight and disposable on the ground" (NASA report, 1984). It has been proposed that at a later stage in planetary explorations the materials may have to be incinerated by a radio-frequency plasma. As for any other invention of civilization, it may be necessary to accompany toilet fixtures with a sheet of instructions on their use. On a Japanese research vessel, I found the following notice: "After flashing, return handle to horizonal position." It was accompanied by the injunction, common in toilets of airplanes and public offices: "Put no matter other than toilet paper into the basin," which, if observed literally, would seem to exclude most of the very materials for which the facilities were designed. Evidently, the educating of infants in the use of such facilities as "totty-potties" is no mean matter: a videotape giving detailed instructions, purportedly under the auspices of the American Medical Association, is currently on sale in North America.

Here is another example of such instructions, in pidgin English, from Papua New Guinea:

> TOK SAVE LONG YUSIM TOILET. DISPELA TOILET EM I KAMAP DOTI STRET. SAPOS YU USIM TOILET, YU MAS HOLIM HANDOL BILONG WARA NA WEIT LONG WARA BAI PINIS. LUKATIM TOILET BILONG YUMI PLANTI. MAN OL SAVE USIM. TENKU! BOS.

Many porcelain or earthenware chamber pots can be regarded as works of art. The British Transport Commission is reputed to have an especially fine collection. Some are elegantly decorated, usually around the outside; others have been devised to play music when sat upon. (One of the happier memories of my student life at Cambridge was sitting on the Backs one warm June evening, after the finals were all over, listening to Elizabethan madrigals sung from punts moored beside King's College bridge. As night fell, the boats were untethered and allowed to drift downstream, followed by a bobbing chain of floating potties carrying lighted candles that glinted along the Cam, which had been thoughtfully provided by pranksters. Gibbons's "Silver Swan" never sounded so sweet. Many years earlier, Queen Victoria had noted toilet paper floating down that same river, and had not been amused.)

The Royal Navy had—perhaps still has—

potties of different qualities and designs, some with appropriate armorial bearings, for use by personnel of different ranks. However, with the advance of domestic plumbing these vessels have tended to fall out of favor and, in falling, to lose their handles or to fragment into shards still occasionally found on rubbish heaps. For popular use a larger, portable W.C. was advertised by one Hawkins back in 1824, many years after the introduction of the Pope's stercoraceous chair mentioned earlier. In a story by Evelyn Waugh, an eccentric army officer insisted on having one such item, which he called a "thunder box," carried with him from one campaign to another. Today's portable versions, bearing such names as "portaloo" or "totty-potty," are widely found at county fairs and state parks, at building sites and allotments, and along many highways and country roads, where they often provide much needed relief. At a recent Native American powwow at Crow Fair, Montana, there were 150 of these so-called (in paleface parlance) comfort stations.

9

SEWAGE

There is now evidence that the flushing of toilets was in vogue as long ago as the Minoan civilization in Crete—at least in the Palace of Knossos—and the middle dynasties of Egypt. Today the disposal of our daily production of feces—several thousand tons in every major city, mixed with urine, paper, condoms, tampons, *aborti,* and such oddments as toothbrushes, spoons, watches, and spectacles, not lost but gone before—confronts civilized life with one of its main problems when it all has to be flushed away with billions of gallons of water, through pipes up to 3 meters in diameter, "down to a sunless sea." Not necessarily directly to the sea, of course; rivers have long been our natural sewers. It is estimated that, even today, some 230 million gallons of raw sewage enter the Ganges daily, with inestimable detriment to those who use the waters for their religious or sanitary ablutions.

Inland towns used to use natural water

courses, which lost their fish populations as they became increasingly feculant. For instance, London's Fleet Street, renowned for its torrents of journalism, takes its name from the river Fleet, which was an open, noisome sewer for centuries until it was finally covered over in the mid–nineteenth century. Around that time, such sewers carried over nine million cubic feet of sludge into the Thames every year. (Now, doubtless, the quantity must be many times as big.) The West Bourne, in Chelsea, bore away what *The Spectator* called "a more or less concentrated solution of native guano." When there was no natural source of running water, as along Edinburgh's old High Street ditches, the smell of accumulated human sewage was even worse (especially on the Sabbath, when strict observance of religious principles precluded its daily clearance). In other countries today, sewage disposal still poses major problems, especially for urban areas. It has been estimated that of the 130 million liters of sewage produced daily by the citizens of Surat, India, less than half is adequately dealt with. One wonders what happens to the rest. In Hong Kong, pig manure, along with other sewage, is channeled into nearby mangrove swamps, where it serves to fertilize the water and thereby enrich the local flora and fauna.

At the National University of Singapore, an island country now justly renowned for its cleanliness, one is liable to a fine of S$150 if one fails to flush a toilet after use. Singapore, blessed with equatorial rainfall, rarely lacks for water, but not every country is so fortunate. In most modern cities, water consumption for this purpose is expensive, and various devices and expedients have been developed to reduce it. In Japan, where today still only 37 percent of the homes are connected to sewers, many of the modern flush cisterns are equipped with "small flush" and "large flush" levers, the latter being used only when one wants to send away solid components. In our semidesert environment in San Diego, California, the city recently gave us $75 for each old toilet that we replaced with a well-designed and more economical Kohler model (the annual sales of which reach about $1.8 billion!), in which the first flush of water impels the solids down to the sewer, and only later is the bowl briefly sloshed down from the top. (The discarded toilets, thousands and thousands of them on our local dump, have been broken up, eventually to be crushed and presumably used for some constructive purpose.) The Kohler system is so effective that we now need to use only 1.6 gallons per flush instead of the 5 or 6 gallons we used before. In Singapore the legal maximum is at present 1.1 imperial gallons per flush. As water supplies become limited, the expenses involved in their use for sanitation increase quite considerably. I note that in my 1997 San Diego city taxes I paid about as much for water (in) as for sewage disposal (out).

These days the collected outflows of many inland cities are passed along to sewage farms, where they are strained, skimmed, settled, fermented, and filtered to render both solid and liquid components as medically, aesthetically, and olfactorily inoffensive as is practically feasible and economically possible. This is not an unmitigated blessing. Abby Rockefeller, who manufactures composting toilets, wrote in her preface to Sabbath and Hall's book *End Product,* "As the missionary bibles once spread light to savage hearths around the globe, the porcelain john will soon spread the Word of Technology to every home of the Great Unwashed." Nevertheless, she claims, "it is clear that the flush toilet must be abolished as a threat to our entire civilization. . . . We must foster technologies for dealing with 'wastes' that make detours in the natural cycle, not irreversible departures." Of course, they are not really irreversible, though they may be inefficient. My late colleague John Isaacs opined that "a major part of the ocean's business [is] converting wastes into living creatures," and pointed out that fisheries' "production in the North Sea has doubled over the past two decades precisely because more and more sewage keeps pouring into it."

Some years ago, the Canadian authorities denied Queen Elizabeth II permission to take her royal yacht up the St. Lawrence Seaway to Montreal because the ship had not been properly equipped with sewage-holding tanks for the royal flush. One wonders how many of the smaller pleasure boats of her loyal subjects could pass this sanitary test. The usual indication of fecal contamination in water is the demonstrable presence of *Escherichia coli,* a ubiquitous bacterial denizen of human guts (White et al., 1996), as well as certain chemicals such as coprostanol. However, not only humans harbor *E. coli.* Its presence in the shore water around the coast of La Jolla, California, where I live now, has been attributed to the excretions of seals, thereby exonerating local authorities responsible for our system of sewage disposal, where leaks had been suspected.

In a couple of ecologically sensitive areas that I have visited in Finland and Alaska, human excrements must not be flushed away for fear of complicating the nitrogen balance of the habitat: instead, they have to be salted down with quicklime and eventually transported to receiving sites elsewhere. From at least some of the inland research stations in Antarctica, such wastes are routinely shipped to South Africa for disposal.

However, in most homes and public establishments (even prisons, unlike those infamous noisome cells at Newgate Gaol and similar institutions of earlier times) in Europe and North America, nowadays one can usually flush away fecal matter relatively unobtrusively. This sometimes makes things difficult for, say, spies anxious to find out about the state of health of public personages. The plumbing system of a Norwegian hotel where Communist Party General Secretary Brezhnev once stayed had to be specially modified, according to some reports, to enable secret agents to obtain the fecal data they felt they needed for diagnostic purposes.

10

TOILET PAPERS AND OTHER ABSTERGENTS

In the British Army we generally called it "bumf." For many years, every sheet of toilet paper supplied in British government institutions was clearly marked O.H.M.S., or more fully ON HIS (or HER) MAJESTY'S SERVICE. (British government employees use some twenty thousand rolls daily; one trusts that most are deployed on the job and relatively few are stolen.) The transparency and glossy texture of the paper were a constant source of wonderment to foreign visitors, a few of whom have used it for airmail stationery, for which it seems eminently suited. In a survey carried out by Consumers Union some years ago, these types of toilet paper were given the highest rating for impenetrability, which is perhaps not surprising since they were issued by H.M. Government. (I have been unable to confirm a story that the Admiralty had once issued toilet paper bearing, on each sheet, Lord Nelson's exhortation to his fleet at Trafalgar: "England expects every man to do his duty.") In England today, toilet rolls are about an inch narrower than their American counterparts, a reflection more of economic than of anatomical differences, necessitating different sizes of dispensers. The dimensions are rarely spelled out on the packages: however, Uzumaru in Japan comes in rolls 65 meters long and 114 millimeters wide, according to specifications on the outside wrapping. (Incidentally, has anyone found a good use for the denuded cardboard cylinders inside the rolls?) At a public facility by the ferry terminal between Nantong and Shanghai, I was recently allotted a single sheet, some 25 by 30 centimeters, presumably to be subdivided however I wished.

In Singapore, shortly after its political independence from Britain was signaled by the manufacture of indigenous toilet paper, the native sheets showed an inconvenient tendency to tear longitudinally rather than transversely, as they should, of course, if properly perforated. Although perforations, as an aid to tearing, were first used for separating postage stamps in 1854, it apparently took another forty years before they were introduced, as a British patent, for toilet rolls.

Fecal textures are of considerable concern to manufacturers of toilet paper, an industry with an annual turnover of several billions of dollars in North America alone, and much scientific research is needed to aid manufacturers in balancing abstergency with ultimate disposability. The eminent Oxford mathematician Sir Roger Penrose recently devised a pattern of stars embossed on toilet paper that somehow makes it feel softer. (*Per astra suaviter,* Romans might have said.) A current lawsuit between the firms of Pentaplex and Kimberly-Clark turns largely on the proprietary use of Penrose's original design. A few years ago, when one was (falsely) led to believe that chlorophyll could be an effective deodorant, some toilet papers were tinted a delicate shade of green. I have not seen any of that kind for some time, and suppose it went out of fashion along with various other "uses" of chlorophyll. Other tints may be employed, and further variety may be conferred by the addition of inscriptions. Years ago, on a visit to an uncle, my little brother once remained for an unaccountably long time secluded on the toilet, and was eventually found, amid a swirl of paper, diligently reading the mottoes inscribed on each sheet. (Horace's poems, which have been printed on other rolls, might not have appealed to my brother at that early age.) Even more time-consuming would be the full employment of a "novelty" toilet paper bearing a separate crossword puzzle on each sheet, although the texture of the paper is obviously unsuited to ballpoint pens or to any but the softest lead pencils. At a private concert, I was once privileged to use paper bearing on each sheet the music of a passage from a Beethoven symphony. And I learned that Madison, Wisconsin, boasts a special museum with more than three thousand rolls of unusual toilet papers, though perhaps they are not all different.

Based on his experiences in Venezuela, William Beebe (1949) wrote:

> Whether or not due to the comparative scarcity of their rightful literary fodder, the bookworms of Maracay seemed to flourish on less intellectual food. I found that an enthusiastically ravenous family had for long enjoyed whatever vitamins existed in the bathroom stationery. As I unrolled it day by day there came to light what looked like an unending scroll of old-fashioned pianola music. Three members of the family seemed motivated by steady, conservative, straightforward burrowings, and the chord of their resultant rhythmic efforts repeated itself endlessly day after day. . . . If I had only kept my old melodeonlike pianola machine, who knows what Folk Songs of the Dermestidae I might have evoked; possibly tissue transcriptions of some entomological symphony of Songs in the Bath!

On the same lines, perhaps, I might mention that some Japanese toilet-roll dispensers are equipped with music boxes, which play a tinkling tune, such as the Scottish folksong "Coming Through the Rye," presumably to keep people in the rest of the building informed that the toilet may soon be vacant again.

As a supplement to toilet paper, there is a device to be found in some Turkish hotels, where toilets are equipped with a little copper pipe that can serve to squirt water over or into appropriate nether regions of the body. Some usage of paper may be saved thereby, but one still needs a sheet or two to dry the areas in question afterward.

In medieval times, when paper was less abundant than it is at present, many people had to resort to whatever came to hand, such as hayballs (which Reynolds [1943] refers to as "mempiria," though I cannot find this word in any Latin dictionary) or little wooden sticks (which the Japanese called *sutegi*). Ladies in the court of Louis XIV were reported to use wool or lace, while the fictional Gargantua experimented with a wide variety of other articles, including live geese. In rural America, *faute de mieux,* corncobs have served the same purpose, though within the last century their use has been largely replaced by mail-order catalogs. The cult leader Jim Jones suggested irreverently that pages of the Holy Bible be used for this purpose, but he was crazy. Abstergents are needed in the field, as elsewhere: as the parson comments alliteratively in Chaucer's *Canterbury Tales,* one wouldn't want clean sheep to be tended by a "shitty shepherd."

Warfare, in this as in previous centuries, involves not only housing, moving, and feeding troops, but also making adequate arrangements for their other natural needs. In the desert, Muslims, as some of their holy writ instructs (I have been unable to confirm this on consultation with a scholarly Muslim friend) are supposed to use at least three stones. However, under no circumstances should stones be dumped into toilet facilities. Unfortunately, this was not made clear to the natives of a Somali village where medical missionaries, concerned about the local lack of sanitary facilities, arranged for the construction of a simple concrete toilet. They were dismayed to find, when they returned a year later, that the installation had been completely clogged by a moraine. From Zen Buddhist aphorisms one may guess that pottery shards could be put to a similar use. Muslims and Hindus agree that such operations must be carried out solely with the left hand, and followed by ritual ablutions. The more strictly these sanitary rules are observed, the less chances one runs of recycling parasites. It has been reported that higher-caste Bengalis, who may have more time and better facilities for such ablutions, tend to exhibit correspondingly lower incidences of hookworm and other parasitic infestations.

In recent years, not infrequently there are shortages of toilet paper and other essentials, exacerbated by wars and transportation problems. There was a "run" on paper in American shops when, in 1974, a projected strike by lumbermen threatened the supply of this estimable commodity. Anyone who has suffered from a shortage or rationing of toilet paper will attest to its value in civilized life today. Charles Mackay, who in *Extraordinary Popular Delusions* (1841) described it as "the most ignoble use to which paper can be applied," should nevertheless have acknowledged this fact.

I was told some years ago, on a visit to the then Soviet Union, that telephone books were not put out beside public telephones for fear that they would be stolen and used as sources of toilet paper. However, in my experience old newspapers and telephone-book pages don't seem to do the job as well as the proper stuff. Russian toilet paper is rarely of the high quality that one generally finds in Western Europe (the "softness" of the currency notwithstanding): at least one French bank in Moscow arranges to have toilet paper for its staff (and clients?) specially shipped from Paris. On the wall of a National Science Research Center (C.N.R.S.) laboratory in Marseilles I saw a charming poster illustrating a naked infant sitting on a chamber pot, with a caption that can be translated: "Don't consider any job finished until all the paperwork has been accomplished." This use of paper distinguishes man from animals. Thirty years ago I noted, in parks and roadsides of arid areas in northwest Mexico, occasional deposits of human feces, distinguishable from those of dogs by being crowned with paper.

Besides its primary function, "loo paper" (as it is familiarly called in Britain) has been put to many other uses, chiefly as inexpensive decoration for children's parties or poor people's weddings, and as wrappings to protect or conceal objects of particular fragility or value. Among the most notable of these were some fossil bones of our putative ancestor *Australopithecus* carried surreptitiously from London to Cambridge by the eminent physical anthropologist Phillip Tobias.

Children who are not toilet-trained need diapers, as many as ten per child per day; millions and millions are needed daily. Every year Procter & Gamble alone sells some several billion dollars' worth. There is much debate as to which material is more environmentally appropriate, washable flannel cloth—involving more work —or disposable paper—requiring more money.

Forest conservationists, garbage disposers, and purveyors of washing machines and detergents favor the former alternative, while people concerned with effluent-water quality, along with cotton growers and paper manufacturers, favor the latter. Basic energy costs are even harder to balance. Having been brought up in a small and often stuffy flat with three younger siblings, I would have greatly preferred that my mother use the disposable variety, but alas it had not been invented in the 1920s. She had to boil the cloths in water plus a handful of washing soda in a cast-iron caldron or hopper, heated over a coke stove and stirred with a wooden stick or "dolly." After rinsing the diapers, she wrung them out or squeezed them through a roller-wringer, and then pinned them in the garden on a clothesline propped up to take the extra weight. In fine weather (all too infrequent in London), they dried, but when it rained they had to be taken in and draped over a rail in front of the gas fire (those were the days before central heating), where their humidity added to the already somewhat noisome fug in the flat. On dry winter days outdoors, they sometimes froze solid, and tended to break if roughly handled. There might have been diaper services at the time, but I doubt it: anyway, we could not have afforded such a luxury.

11

NUTRITIONAL VALUES

All kinds of dung, though consisting of materials excreted by one kind of animal, may nevertheless retain considerable value for others. It has been calculated that human feces retain some 8 percent of the calorific value of the food originally ingested, but this of course depends on the diet of the excreter. A Chinese farmer has estimated that, assessed fecally, one Englishman is equivalent to three pigs or eight Portuguese, while the quality of "night soil" from wealthy areas of Japan is supposed to be appreciably higher than that from poorer districts. In Jonathan Swift's account of Gulliver's travels, he described the activities of an old student of the Academy of Lagado whose task it was to fractionate excrements and retrieve whatever nutritional elements they retained. But generally it is some other species that takes advantage of this source of nourishment.

Chicken manure, which in the United States today sells for $15 to $45 a ton, is widely used as a nitrogenous supplement in cattle feed. It is usually allowed to ferment for ten days or so, a self-heating process that kills many potentially transmissible pathogens. The droppings of king vultures, which feed largely on carrion, retain enough nourishment to constitute delectable morsels for peccaries that forage under the birds' nests. The feces of carnivores, though one would expect them to be richer in nutrients than those of horses and cattle, attract relatively few kinds of dung beetles, and their decomposition is largely attributable to bacteria. But in pastures of temperate lands, the undigested matter in cowpats provides a welcome source of food for all kinds of insects. Some sixteen of the commonest types in Britain are tabulated by Putman (1983), with discussions of their several preferences. Putman has quoted extensively from a 1978 Oxford University doctoral thesis on the subject by P. A. Denholm-Young, which contains much fascinating information.

Cows, like deer and sheep, are ruminants: they chew the cud. This means that much of the relatively undigestible matter in their food, notably cellulose, is held to ferment in one of their several stomachs. Perhaps this accounts for the fact that cow manure is squashier than, say, horse droppings. In horses, which have no rumen, late digestion takes place in the lower regions of the gut. Nevertheless, their droppings still retain a lot of relatively firm material such as chewed straw and even intact seeds, which are among the main items of diet of the ubiquitous, unsqueamish sparrows and rarer South American macaws.

Small scats of animals dry out faster than larger ones, of course, and thereby become less attractive to such insects as dung beetles. The adult beetles generally slurp up the rich juices through their mandibular teeth, leaving the less nutritious solids to be nibbled by their grubs. The richness of dung diets may be exemplified by grubs of the horn fly *Haematobia irritans,* which in warm climates, eating nothing else, can complete their whole larval development in less than a week. Up to one hundred species of insects, including even termites (though most species seem to prefer a diet of wood), have been found in a single cowpat. At various times, as the turds age, different species predominate in a more or less orderly succession until the pats disappear. (In England this takes one to five months, depending on the warmth of the season.) Without insects to promote their disintegration, and left solely to the exigencies of the weather, the pats might persist for twice as long.

Among the best-known species of insects to feed on pats are dung flies, of which *Scatophaga stercoraria* (the "scat eater") is the best studied. On a fine summer day, the social behavior of these large, yellow predatory insects is easily observed. After emerging from their puparia, the male flies stand around on soft, new cowpats, sometimes jostling for space at the edges, and try to hop onto passing females, with which they mate. Large males mate for about twenty-five minutes; small ones tend to take twice as long. These copulating pairs of hairy flies are doubtless familiar to every countryside pat-watcher, even those unaware of their amorous activities. Their larval grubs, which feed on the dung, constitute an important item of diet for many kinds of farmland birds that excavate them from the pats. (In Britain two thirds of cowpats are at least partially pecked apart by birds before they finally disintegrate.) Other grubs may succumb to persistent residues of vermifuges such as Ivermectin, administered these days to cattle to cure them of worm infestations.

Ordinarily, though, cowpats are favorable habitats not only for insects but also for worms, notably nematodes. One scientist reported finding 13,440 such worms, belonging to 34 species, in only 18 pats. (According to folklore in England, perhaps also elsewhere, horse hairs that fall into manure piles may develop into snakes.) And of course cowpats are full of microbes of all sorts: a gram may contain more than a million yeast cells, as well as 10,000 or 100,000 times as many assorted bacteria. Fresh cowpats are the best places to look for *Caryophanon,* a most unusual and relatively enormous kind of bacterium found almost nowhere else in nature except inside cows.

Bourke (1968) reported that certain natives of Florida—presumably pre-Columbian—used to eat the droppings of some deer, as do starving Aleuts and Inuits when there is little else but reindeer pellets to alleviate hunger. Coots may eat goose droppings. In the Antarctic, gull-like birds called sheathbills (*Chionis*) enjoy a diet of seal feces and afterbirths, while at the other pole, around Wrangel Island off the north coast of Siberia, phalaropes snack on walrus feces and, farther west, Eskimo dogs stave off the pangs of hunger by consuming whatever dung they can find, along with skins, bones, and other offal. Many of our domestic dogs quite happily consume cat droppings, which, because cat diets tend to be rich in proteins, usually contain more residual nutritious components than their own. And although they may have spent a week or two in transit, there is still enough "goodness" in the droppings of burrowing gopher tortoises, which live in holes on riverbanks in Florida, to nourish an underground population of mole crickets.

Under water, hippos and manatees are often followed by schools of hungry fish, waiting to garner what, for want of a better word, we might call handouts, and of course many larger fish have their followers similarly waiting to snap up the sporadic benefices. There are some species of fish that choose to spend their lives hanging around the cloaca of a whale shark, just in order to be among the first to snatch fecal morsels as they emerge. Some herbivorous fish that follow carnivorous species around can considerably enrich their diet in this way.

The late but eminent ecologist G. Evelyn Hutchinson, in his exhaustive study of guano (1950), has reported that some bat species seem to be more ecclesiophilous than cavernicolous, though in most church belfries the accumulation of bat droppings does not constitute a serious problem (perhaps because bats generally dislike loud noises). Accumulations of bat droppings may constitute a groaning board for the delectation of the hordes of giant golden cockroaches, and even a few kinds of caterpillars, which throng many bat caves, especially in the tropics. Some birds' nests harbor coprophilous caterpillars, and bird droppings are among the main dietary components of many species of daddy longlegs, or phalangids. It has been reported that on Mount Elgon (between Kenya and Uganda) elephants that go into the caves to eat lumps of zeolite, as mineral supplements to their diet, seem to prefer those flavored with bat shit.

The feces of invertebrates and their disintegration in nature have been less studied than those of vertebrates. Some caddis-fly droppings decay, due to bacterial action, in two to three weeks, whereas thosc of millipedes may persist for several months. One marine biologist (Turner and Ferrante, 1979) has devoted at least ten years of research in North Carolina largely to studies of copepod feces. He has discovered that, among the main components of zooplankton, copepods produce fecal pellets which, like those of many bird nestlings, are encapsulated in a special organic coating of unknown function. Produced at rates of five to seventy per day, many of these pellets stay intact for a few days. Most fall through the water at dignified rates of 1 to 3 meters per hour, although if bubbles form in them as a consequence of anaerobic bacterial fermentation, they may even float for a while. After their bacterial disintegration, the comminuted contents fall even more slowly, perhaps taking years to reach the ocean depths. But while they remain intact, the pellets may be consumed by other shrimplike plankton, or by midwater worms like

Poeobius, for which the partially digested food provides a welcome dietary supplement. Fecal pellets of the brine shrimp *Artemia*—extruded at the very end of a long tail, perhaps to minimize their reentry into the intake part at the front end of this filter-feeding crustacean—are neatly encapsulated like sausages in a cylindrical skin of chitin. When the pellets sink to the bottom of brine ponds where *Artemia* abound, they may constitute a major component of the diet of clams living in the mud.

Incidentally, the presence of so many diatom skeletons in the pellets of copepods and other zooplankton suggests a theory to account for their mineralization. The silica walls of diatoms—fragile, often spiny, and no more digestible than broken glass—cannot fail to have a constipating effect on the little animals that eat them. This might be expected to confer a selective advantage on the diatoms—not to the cells already eaten, of course, but to their genetically identical siblings still living in the plankton. (Maybe this could have contributed to the success of diatoms in evolution.)

The sweetest anal excretions in nature are probably those of aphids, as many gardeners (both ant and human) know well. They consist mainly of a concentrated solution of sugar, hence the endearing term "honeydew." Many species of ant go to a lot of trouble to corral and protect herds of aphids, which reciprocate for these attentions by providing liquid refreshment for their insect shepherds.

Among the more familiar kinds of droppings are those of domestic mice and rats, which live with us and share our foods. In the Second World War, when wheat was rationed in England, conditions in granaries and silos were so unsanitary that mouse droppings were common among stored grains. A certain number of these feces was legally permitted per pound of seed to be milled and used for making bread (in which process, one assumes, they were sterilized and thereby rendered harmless). Unknowingly, most of us who ate bread in England during that period must have consumed quite a lot of rodent droppings, and I have no doubt that in other countries the contaminations were even higher.

Biblical scholars are still unsure whether the food substitute sold during a period of famine at five shillings a pint was really pigeon droppings, as stated in 2 Kings 6:25, or whether it was some sort of porridge which had been given this appellation as a vulgar joke. (A plant in the soapwort family is called, in Arabic, "sparrow dung.") A similar sense of humor is exhibited in Thailand, where little green chili peppers are called (in Thai) "mouse droppings."

12

DUNG BEETLES

Most kinds of mammalian excreta are soon recycled, generally by insects. Remarkably, "it was not until 1960 that the first work was undertaken in analysis of the actual processes involved in decay" of dung (Putman, 1983). In temperate climes, earthworms are perhaps the chief agents responsible for the breakdown of sheep droppings, while rabbit pellets are a favorite food of minotaur beetles. The best known of all coprophiles are the dung beetles, which the French call *fouilles-merde*. Their dung-rolling activities are epitomized in the logo of the New York Entomological Society, where their more usual burden has been symbolically replaced by a globe representing the world. Although many kinds of dung beetles abound in tropical regions, only three genera are recorded in temperate zones, and only one species, *Geotrupes spiniger*, is really significant here. Feces of humans and other omnivores, which are generally richer in nutrients than those of herbivores like horses and cattle, attract few dung beetles, which mostly seem to prefer the droppings of vegetarians. A few dung beetles specialize in human excreta, though this material is rarely available to them nowadays in civilized countries; most exploit the droppings of large herbivores such as cattle. Generally, each species of mammal in nature tends to nourish its own kinds of dung beetles.

There are thousands of species, in a wide range of sizes, all devoting much of their lives to gathering dung from all sorts of animals, ranging from mice to elephants, which—sometimes even associated with their dung balls—first appeared among fossils of the Mesozoic era. They merit a special book of their own, and in fact such a volume has been recently compiled by Ikka Hanski and Yves Cambefort (1991). These scientists clearly share an unbounded enthusiasm for their fascinating subject. And no wonder: there are 5,000 species of Scarabaeids, 1,850 Aphodiids, and 150 Geotrupes, all capable of dealing with various kinds of dung in their own special ways. Hanski (who in 1989 obtained a doctorate for a thesis on dung-beetle ecology) seems to have developed an almost protective feeling toward animal droppings, which, he says, "unlike mushrooms . . . do not defend themselves and may be burrowed into, buried, or rolled away by beetles, each according to her own special way of life and the prospective needs of her progeny."

Time-lapse cinematographic studies made in East Africa have shown how quickly even the biggest heaps of pachyderm turds (obviously among the largest such droppings, and the ones most studied) can be divided up and trundled away. Some sixteen thousand beetles that thronged to a three-pound heap of elephant dung completely dispersed it within a couple of hours. (No wonder the ancient Egyptians regarded them as models of diligence!) But that occurred under moist conditions. In a dry season, a single bolus may persist for four months, and it may take a couple of years to disperse a complete elephant dung pile. The beetles clearly treat dung as a valuable commodity, sometimes fighting over possession of choice bits; males of certain species may present prospective mates with small pellets as nuptial offerings. From each fresh pile some beetles remove chunks as quickly as possible to take them out of the reach of competitors, while others may stay to guard their chosen bits, or fly off to seek fresher piles. (Incidentally, they may take with them mites and nematode worms that share their taste for dung. Some of the mites pay for the rides, as it were, by eating the eggs of flies that had reached the dung before the beetles got there.)

Some beetles simply drag away portions of a size and general shape suitable for this purpose, but the most industrious scarabs characteristically remodel the dung into neat spheres to facilitate its transportation. A hundred years ago, the naturalist Henri Fabre commented on the skill of his local *Sisiphus schaefferi*, which somehow manages to sculpt the spheres even before they are detached from their matrix for transportation. Like other talented species, using their back legs like the forks holding the wheels of a bicycle, the beetles pivot their spheres on an invisible axle, thereby operating the nearest thing in the natural world to a true wheel. They trundle backward, which must make steering toward their prepared dung-burial sites even more difficult. Whereas some species, like *Sisyphus,* do their rolling singly, others work in pairs, and in not a few cases one sex (guess which!) does the pushing while the other rides on top. (For millennia, at least since the times of the pharaohs, the diligence of dung beetles has been held as exemplary, perhaps to encourage slaves lugging two-ton blocks of stone for the construction of the pyramids. Although Egyptologists assure us that the spheres carried by scarabs, in carvings or inscriptions, represent the sun, I favor a much more down-to-earth interpretation.) In carefully selected sites, they bury them. (A Sri Lankan scientist reported once seeing a beetle roll his hard-earned treasure into a crab hole by mistake, so incensing the resident that a fight ensued. The unfortunate insect escaped with his life but without his booty.) Normally an egg is laid in the middle of each sphere, where the larva that eventually hatches finds both shelter and dinner.

The hard carapaces of these beetles can even persist as fossils, the commonest being those of *Aphodius holdereri,* from the last ice age. This species is now almost extinct except in the high plateaus of Tibet, one of the few places where, one may suppose, they can indulge their special preference for yak droppings.

Some species, having cached a suitable quantity of dung in a burrow especially excavated for the purpose, go on to rework the material underground, fabricating even larger spheres to nourish their prospective young, and carefully covering them with clay to conserve their moisture. African ratels (sometimes called honey badgers, though their diet includes other items besides honey) sniff out and dig up such balls to snack on the delectable grubs inside.

One of the causes of the fly pestilences that in warmer months used to plague parts of Australia, after the introduction of cattle in 1788, was the absence of native species of dung beetle, able and willing to dispose of the daily tonnages of cow manure. Since most of the native Australian species are accustomed to dealing solely with the feces of marsupials, conditions in that continent became increasingly unpleasant. It was asserted that in some parts the pats "almost merged into one continuous habitat." It was only after the 1970s, when an extensive Commonwealth Scientific and Industrial Research Organization program was started in which some twenty selected species of African dung beetles were introduced, that Australians managed to ameliorate the problem. In some places, the populations of dung flies, which had become so much of a nuisance in earlier years, were reduced by as much as 90 percent, a partially successful biocontrol. Unfortunately for these insects, and indirectly for ourselves, cattle vermifuges like those mentioned in Chapter 11, which finish up in the feces, may wipe out the dung beetles along with the parasites.

Two informative television programs on the subject, both British productions, were *The Wonderful World of Dung,* dealing with animal excretion in general, and *Tombs Under Aruba,* beautifully illustrating the activities of dung beetles.

13

REFECTION, TRANSFECTION, AND DISSEMINATION

Eating feces is called "coprophagy"; this subject has been reviewed by Hörnicke and Björnhag (1979). Eating their own feces is a natural phenomenon among certain animals: it has been called "caecotrophy." For optimum growth, rats in captivity need to practice it, especially if their diets are deficient in vitamins. The practice is not unusual among certain other rodents, rabbits, and their kin which, though they live on grasses and other cellulosic vegetable matter, lack the long and elaborate intestinal systems of ruminants, which might impede their mobility. So they effectively double the lengths of their guts by sending everything through twice. As a result of what is called "colonic sorting," the nighttime feces of rabbits are soft and black, consisting of partly digested grasses and other leaves, and are nibbled straight from the anus as a kind of breakfast. Like the pellets of many nestling birds they are held together by a special mucus that keeps them intact (see page 50, Thacker and Brandt, 1955). They are swallowed whole, and sometimes retained intact in the stomach for several hours before the contents are recycled. The fermentative bacteria in their intestines, which help to soften much of the ingested vegetable matter, convert some of it to short-chain fatty acids with characteristic smells, and it has been noted that rabbits raised under bacteria-free conditions rarely refecate, presumably because their droppings lack the appealing odor of those acids. Pikas, which are related to rabbits, sometimes stick their soft feces (which may be only a couple of millimeters wide) on rocks and lunch on them later. Daytime rabbit droppings, drier because most of their "goodness" (two thirds of the protein) has been digested and absorbed, are the kinds more familiar to us in fields around the countryside. (To their credit, rabbits do not ordinarily foul their burrows.)

Rats and mice refecate, too, as do beavers, ground squirrels, guinea pigs, chinchillas, and even a few species of lemur. In some cases, recycled droppings may constitute up to 25 percent of their total diet. This phenomenon tends to complicate nutritional studies carried out in laboratories evaluating dietary components; the animals may have to be fitted with special collars to prevent refection. Laboratory rats, deprived of the privilege of eating their own feces, may ultimately come down with deficiencies of vitamins K and B_{12}. Mouse droppings, especially those produced in the early morning, are found to be enormously enriched in vitamin B_{12} as well as appreciable amounts of at least five other vitamins made by the intestinal bacteria, which could certainly account for some of their enhanced nutritive value (at least to mice).

Such a process of recycling occurs among various other leaf-eating animals, including reptiles and mammals of various orders, where it can be regarded as nutritionally equivalent to chewing the cud. Colts, although they are not technically ruminants, sometimes eat their mother's droppings. Especially in times of drought, elephants have been observed eating their own dung, perhaps as much for its moisture content as for its residual nutrient value; a few vegetarian lizards, notably the desert iguana *Diplosaurus,* are reported to do likewise. Some shrews, although carnivores, are also refective, perhaps because food passes through their short guts to their cloacae too fast for efficient utilization of proteins and other nutrients. These small animals have what might be called a "rapid transit system": in the smaller species, the passage from food to feces takes as little as one hour. (For comparison, a passage time of about sixty hours is par for adult humans. Based on analyses of hundreds of self-made observations on defecation, Probert et al. [1995] found that the usual transit time was about two and a half days, but was reduced to two days in men who were heavy drinkers.) Agricultural researchers have carried out experiments on the nutritive value of pig droppings when fed back to pigs. Scientists trying to propagate caged colonies of the rare snail *Partula* discovered that they tended to die out if the cages were kept too clean and the animals were prevented from snacking on their own fecal matter from time to time.

Seeds of the guanacaste tree may be distributed first by horses and then by pocket mice, which gather seeds from the horse droppings and carry them away, sometimes for considerable distances, to store in their underground caches for winter use. For some reason, among the most fragrant coffees of Java are reputed to be those made by roasting beans that have passed through the intestinal tracts of civets. However, it is hard to see how this process could be used on an industrial scale.

According to the eighteenth-century Mexican historian Clavijero, natives of western Mexico used to carry out somewhat similar activities, probably because in winter they must have been particularly short of food of any sort. (Understandably, the Mexican goddess Tlazolteotl, the eater of ordure, was not regarded particularly highly.) After eating *pitaja* (fruits of the cactus *Machaerocereus gummosus*), the Indians would defecate in specific locations near their encampments where later the small indigestible seeds could be sorted out, dried, roasted, and ground to flour to supplement their meager diet.

Other animals, including many songbirds and rodents, shrews, wild dogs, hyenas, giant pandas, okapis, and antelopes like the Arabian ream, practice a form of transfection, in which the mothers eat the feces of newly born offspring. The mother antelopes may take the feces directly from the anus. This may be a way to reduce evidence of their presence, which might attract predators, but almost certainly it is also because the feces have some nutritional value, and probably also because the act stimulates anal activity in the young. Although naked mole rats, which live in underground burrows of considerable complexity, customarily use special chambers as toilet areas, their young, when they feel the need for a few mouthfuls of fresh feces, usually solicit them directly from passing adults. Transfection occurs among our own close relatives, including the predominantly leaf-eating monkeys, such as langurs, and even gorillas—though understandably coprophagy by humans is unnecessary and indeed deprecated (see *Ubu Roi,* Rabelais, and Isaiah 36:12). Nevertheless, it does occur among small infants and certain mentally deranged adults.

Before they are weaned, baby vampire bats pick up from parental feces the cultures of intestinal bacteria (notably *Aeromonas hydrophila*) that they need to digest their special diet of blood. Some kinds of colonial insects, such as termites, eat the feces of their nest mates for much the same reason; Maeterlinck suggested that by piling the feces in corners and allowing them to dry, *Drepanotermes* somehow renders them more appetizing. A diet of wood scrapings would be quite indigestible to such insects if it were not for the activities of symbiotic protozoa (*Trichonympha,* etc.) and bacteria living in their guts. These microbes have to be handed on from generation to generation by "transfection" or "trophylaxis." For similar reasons, some kinds of flea larvae supplement their diet of fur, feathers, and fluff by eating parental frass.

Most pouched animals, like kangaroos, clean out their pockets before too much dirt accumulates from their joey's excretions, but some female koalas are so remiss in their personal hygiene that the mess in their pouches may accumulate until it suffocates their unfortunate offspring. (Maybe koala pouches, since they open to the rear, are more difficult to clean out than the front-opening kind found in kangaroos.) However, koala babies have to be weaned on a diet that includes some of their mother's excrement, as well as milk, if they are to survive. This is because the adults eat only eucalyptus leaves, and to digest the leaves the koalas have to have large intestinal systems in which leaf fermentation can be effected by special bacteria (as in cow rumens). For the infant koalas, getting the right gut microflora necessitates eating some of their adult parents' excreta, which, as one might guess, smell strongly of eucalyptus oil. Other vegetarians, including rats, beavers, hamsters, and iguanas, do likewise. Maybe goslings do something of the sort, too. (Geese have exceptionally long cecal sacs, which probably serve the same role as mammalian rumens.)

Tomato and fig seeds, which seem to survive passage through the human alimentary tract, are a characteristic component of human feces. When the seeds are discharged eventually into the sea, they constitute an appreciable fraction of the "plankton" samples that can be collected by fine nets in, say, the English Channel, though it is unlikely that any could germinate there. In many sewage treatment plants, the abundance of tomato plants attests to the remarkable viability of such seeds after successfully withstanding human digestive juices.

Some birds are among the primary agents responsible for distributing seeds in their droppings; indeed, the germination of many kinds of seeds may be enhanced by passage through a bird's gut. Darwin sorted seeds from all sorts of birds, from small passerines to eagles, and showed that many of them were still capable of germination. The abundant fig bushes along the banks of the river Don have probably similar origins, while in Java many other kinds of trees and shrubs are disseminated in the droppings of birds, especially fruit pigeons. At certain times of the year, the droppings of cedar waxwings tend to consist almost exclusively of the seeds of rowan (*Sorbus*) or firethorn *(Pyracantha)* berries, of which these birds seem to be particularly fond. Such seeds may take only a half hour to pass through the birds, or even less. Australian mangrove mistletoe fruits have some sort of laxative action that induces the birds that feed on them to excrete their seeds within as little as ten minutes (thereby reducing the distribution range to only a few kilometers). In Europe mistletoe seeds are distributed in this way, sometimes over distances of several miles, by thrushes: in the southwestern United States, the seeds are carried by desert birds, notably *Phainopepla*. The sticky seed coats (the basis for old-fashioned birdlime) retain their adhesive properties even after passage through the bird's gut, and strings of seeds, eventually excreted, have thus a good chance of looping over and sticking to branches if they are dropped over suitable trees.

Nestlings of the rhinoceros hornbill of Thailand have developed the art of shooting their droppings out through the entrance of their nest hole in the trunk of a tree. The seedlings that in due course grow at the foot of the tree provide evidence of the kinds of fruits provided by the parents. (As already mentioned, this projection skill is also exhibited by baby kingfishers.) In some of the larger caves in the Venezuelan mountains, where oilbirds (*Steatornis*) congregate and nest, the seedlings arising from their droppings—a soggy morass of guano—give rise to miniature forests of blanched palms and other plants, which have germinated but can grow no further in the almost complete darkness. In the drier regions of Central America, seeds of the cardon cactus are dispersed in the feces of fruit-eating bats, while in the rain forests natural or manmade clearings are soon repopulated by young plants, including fast-growing saplings of the *Cecropia* tree, arising from seeds in bat droppings. Some 40 percent of the feces of howler monkeys are dropped around their overnight lodgings, and the plants arising from seeds contained in them are correspondingly localized.

By counting the numbers of date stones in coyote droppings found at various distances from the nearest oasis in the Arizona desert, scientists have been able to estimate the areas of coyote foraging ranges. Seeds of another palm, *Washingtonia* (which are considerably smaller than those of the date), as well as wild-plum stones, are distributed over even larger distances by coyotes and foxes in the Anza-Borrego Desert of southern California. On an even bigger scale, the droppings of tapirs disseminate palm seeds in the Amazonian jungles, while on the plains of Africa elephants and impala not only serve to distribute the seeds of *Acacia* plants, but also apparently promote their germination and growth. A recent study has shown that some species of seaweed may be dispersed in the fecal pellets of marine mollusks on the coast of Chile, and doubtless many similar examples could be found elsewhere. Also relevant in this connection is the observation that in giant clams (*Tridacna*) that harbor symbiotic algae, when the algal cells breed faster than they can be conveniently accommodated, the excess are expelled with the feces into the surrounding seas.

Spores of fungi as well as seeds are disseminated in animal droppings. We find the Greek word for feces, *kopros,* in the name of the fungus genus *Coprinus,* which has hundreds of species adapted to various animal dungs. Spores of such edible mushrooms are naturally dispersed after being eaten by animals, in whose guts they survive and by whose excretions they are not only distributed but also fertilized. Kept warm and moist, the feces of many wild animals yield characteristic kinds of *Coprinus,* depending on which animals consume the fruiting bodies. (Although *Coprinus comatus*, the "inky cap" fungus, is extremely good to eat, since humans usually eat it fried, I doubt whether we are major agents for its dispersal.) Spores of the stinkhorn *Phallus impudicus* may be dispersed in the droppings of flies, while squirrels play an important role in the dispersal of truffles. One of the commonest dung fungi found in droppings of all sorts of herbivores (including extinct ones like mammoths) is *Sporormiella;* it produces its spores just under the surface of the pellet until they are ripe enough to pop out.

Campylobacter, a helical microbe common in cowpats and sheep droppings, is sometimes transferred by magpies from pastures to milk bottles left on doorsteps, where it can be a source of infection to humans. Cowpats are also reported to be sources of the dissemination of *Mycobacterium* (which can cause tuberculosis) by starlings and even by badgers, rummaging in them for grubs. (As a consequence, in some regions these charming mammals have been almost exterminated by health-conscious hunters.) A species of yeast that causes a lung disease, histoplasmosis, is also believed to be dispersed by the droppings of various kinds of birds.

Some of the many fungi that thrive in dung produce special antibiotics to suppress growth of bacterial competitors. Other specialized fungal features include an ability to shoot more or less sticky spores away from fecal masses to surrounding grasses, where they may be taken up by the next animal grazer. The dung fungus *Pilobolus* shoots its ripe spore balls toward the light, thereby ensuring that many of them land on grass tufts beyond the edge of the cowpat, where they are more likely to be eaten. In this fashion, the fungus completes its life cycle. (For more fascinating details, see Dix and Webster [1995] and Chapter 14 in this book.)

14

HEALTH AND DISEASE

I'll deal first with purges here, although the subject might be more appropriately discussed in the chapter on myths and legends, since most components of purges tend to be based more on lore than on logic. It may seem natural to think that, since feces are waste matters that generally smell nasty, their elimination should carry away other nasty matters. And so, in almost all epochs and societies, purging has been one of the most popular ways of dealing with bodily ills—popular more with parents and medicine men than among children and patients. It is also true that an overfilled gut can make one feel logy, and purging lightens one's load and may within limits lighten one's step, too. However, many diseases are in the mind, or in organs of the body other than the gut, in parts that cannot be reached by purges, while even those pathogens and parasites that afflict the intestines can only rarely be eliminated by catharsis. Tapeworms remain hooked on, and though billions of bacteria and protozoa may be expelled with feces, billions more remain and the disease goes on. These days, purging is rarely advocated—at least, where I live—except for specific ends, for example, as a means of evacuating the lower gut prior to a colonoscopy.

One hundred bacterial species, in some forty genera, have been identified in human feces; almost all, fortunately, are harmless or even beneficial. A newborn infant starts with none, but quickly picks up a varied bacterial flora, presumably much of it originating from maternal nipples. Within its first three days after birth, the predominant microbes are generally enterobacteria, but the feces of a week-old child may contain 10 to 100 billion bacterial cells per gram. Most of these comprise the species of *Bifidobacterium* that eventually, after several months, characterize the feces of normal adult humans.

Healthy human feces are compacted and accumulated in the colon after much of their water content has been absorbed in the small intestine. They are usually more or less cylindrical, although in a condition known as "irritable bowel syndrome" they may be pelletlike and emerge as small balls or ribbons. They are usually voluntarily expelled under suitable and convenient conditions. Early in this century, Dr. John Harvey Kellogg, a brother of the eminent cerealist Will Keith Kellogg, recommended five bowel movements a day, which generally necessitated frequent colonic irrigation, although most people nowadays would consider this excessive. Repeated voiding in inappropriate places is called "encopresis."

Constipation is common, especially under conditions of tension or stress. In some people it tends to induce headaches, possibly caused by toxins accumulated in the lower gut and transmitted through the bloodstream to the brain. Henry James need not have bothered to tell us about his intestinal problems. Since 1909, when Sir Arthur Hurst published his monograph on the subject, the medical literature has burgeoned: a more recent review lists 599 references (Devroede, 1993). A survey by anonymous questionnaire of five thousand American naval personnel (Sweeney et al., 1993) indicated the following: "When . . . defined as no bowel movement for greater than 3 days, 3.9% . . . are constipated when in their home environment, 6% when aboard ship and 30.2% while in the field. When defined by . . . hard stools, painful defecation, and bleeding . . . the [respective] incidences are 7.2%, 10.4% and 34.1%." One wonders how typical such proportions are among people of other nations, other professions, other circumstances, etc. It has been suggested that Christopher Columbus and his crew might have alleviated constipation, consequent on their generally desiccated diets, by their daily allocation of 2.5 liters of Spanish red wine. It was by terminal constipation, if we are to believe Balzac, that one Sister Petronille strove to achieve canonization, as Saint Catherine of Siena had done in an earlier epoch.

Innumerable remedies, including enemas and

lubricant suppositories of various kinds, have been proposed and compounded to cure human constipation. We are told that in a single year the French king Louis XV was subjected to more than two hundred purges and a comparable number of enemas. (Evidently not all of his time was spent on the throne.) Seeds such as those of psyllium (*Plantago afra*) and chia (*Salvia* spp.), with mucilaginous coatings of xylan or other carbohydrates not subject to breakdown in our digestive systems, are among the most effective. So, too, are remedies based on herbs containing mildly toxic compounds that stimulate the mucous membranes of the gut to exude a lubricant liquid. Alexander Pope may have been referring to such a phenomenon when he wrote (*The Dunciad*, 2:103–4) "Renew'd by ordure's sympathetic force, As oil'd with magic juices for the course . . ." Epsom and Glauber's salts (magnesium and sodium sulfate, respectively) act mainly by increasing osmotic pressure, but perhaps also by irritating the secretory cells of the intestinal lining. Among the oldest and most widely used remedies for constipation is rhubarb, renowned in ancient China and Persia, where it was once extensively grown for trade. Ex-Lax chocolate, however, at least until recently has owed its action to being laced with a synthetic chemical, phenolphthalein. This substance is colorless in acids, but turns mauve in alkaline solutions, sometimes leading to colorful fecal effects and occasionally to amusement, consternation, or panic in those who observe those unusual colorations.

In many animals, alternating conditions of diarrhea and constipation are usually indications of worm infestations. Sabbath and Hall refer us to little-known seventeenth-century texts giving recipes for clysters to be administered to constipated dogs, horses, and oxen. For human constipation, the use of clysters or enemas (both words being of Greek origin) seems preferable to dosing with castor oil, which fell out of favor in the 1930s and 1940s when it was used by Mussolini's thugs in Italy in vain attempts to cure antifascism.

In 1998 a report was widely circulated by E-mail about a tragic event at the zoo in Paderborn, Germany, where a keeper had so successfully treated a constipated elephant with a laxative and an enema that the 100 kilograms of violently expelled feces had felled and suffocated him. I have been assured by authorities at Paderborn, which in fact has no zoo, that this story is wholly unfounded.

As our mothers and many laxative advertisers have long insisted, regular bowel movements, every day or two, are generally prerequisites of good health. In a nematode worm, *Caenorhabditis elegans,* the processes of excretion are controlled by a single pair of nerve cells, one in the head (designated AVL) and one in the tail (DVB), and genetic impairment of either tends to lead to death by constipation. However, for some other animals the system may be quite different and "regularity" may be on quite a different time scale. For instance, a hibernating bear may not excrete for many months, while one of the first acts of an ant-lion adult is to relieve itself of the fecal matter it had accumulated throughout its larval life, a year or so, in its dead-end gut.

Diarrhea is one of the major symptoms of infection by *Salmonella* or *Campylobacter* (or both), all too commonly picked up from uncooked or incompletely cooked chicken or ground beef. The former bacteria are generally associated with contaminant fowl feces, which could be killed by radiation, although, as a U.S.D.A. authority has affirmed, sterilized feces are still feces, and we prefer not to take them with our meals. All sorts of conditions and agents induce diarrhea besides bacterial and amoebic infections, including scurvy, concussion, eating too many green apples, drinking pond water containing toxic blue-green algae, and nervousness—as indicated by the Australian expression "having the shits" or the Yorkshire dialect word "squit." Even something as apparently bland as cow's milk can have a similar effect on those humans (and seals, incidentally) who lack lactase, the digestive enzyme that breaks down lactose. This "milk sugar" is then moved down the gut until it is fermented and produces acids in the colon. The lactase enzyme seems to be genetically absent from many people of African or Oriental descent, including the original natives of North America. Milk offered apparently in good faith by prospective European settlers to Sioux Indians, around the middle of the last century, was not appreciated for this reason.

A medical colleague told me of having probably saved the life of a tourist (visiting a country where diarrhea is rife), who had been misguidedly taking daily what he considered to be prophylactic doses of the drug atropine. If he had continued to do, the pills would probably have constipated him to death. Rather than numbing one's guts with an intestinal narcotic such as atropine or codeine, which effectively slows things down, one can overcome a temporary diarrhea by the more mild expedient of ingesting some kind of emulsified clay or kaolinite. Along the same lines, perhaps, is an ancient Chinese remedy composed largely of earthworm casts—predominantly earthy material—rendered a little more palatable when taken with a yellow alcoholic spirit.

Human bowel movements are subject mainly to reflex reactions, although one Westerner commented on the "phenomenal capacity of the young Bantu to defecate upon request." Like ourselves, in whom this is a well-known response to adrenaline secretion, many mammals (even elephants), birds, and fish may be induced to excrete precipitously when frightened. An ecologist in Malaysia who wished to collect samples of bird droppings, for studies of their intestinal parasites, had only to catch a creature and pop it into a plastic bag for a few moments, generally to be rewarded with a sample adequate for his research purposes. And another colleague of mine, while standing under trees in one of the isolated copses that characterize the Paraguayan pampas, experienced not only vocal but also physical evidence for the presence of howler monkeys overhead. He generously attributed their gut reactions more to fright than to malice. I had a similar experience in Venezuela, but I noted that the fecal pellets were solid, not runny (as I would have expected if the monkeys had been really frightened).

In some tropical regions, diarrhea (or polychasia), caused by *Salmonella, Vibrio,* and other bacteria and protozoa that may infest our digestive systems, is endemic; in poor areas in northeast Thailand, for instance, most human feces are normally fluid. Where sanitary arrangements are inadequate, diarrhea is especially common among children (and, unless they take reasonable precautions, their parents too). It has also smitten so many unwary or improvident tourists that the condition has received special appellations, "Montezuma's revenge" or "la turista" in Mexico, "Delhi belly" in India. Actually, diarrhea should not be regarded as a joking matter. According to estimates recently published by the United Nations, today it still causes the deaths of over three million children annually, a tenth of them in cities of India alone; how many more in rural areas would be hard to estimate, but the numbers would certainly be large. In a recent outbreak in 1993, attributed to *Cryptosporidium* in contaminated drinking water, some 400,000 cases were reported in and around Milwaukee, Wisconsin; fortunately few of them proved fatal. Cholera, one of the primary causes of dysentery, apparently brought to South Africa from India early in the nineteenth century, probably killed more soldiers in the Boer War than did armaments, while in more recent outbreaks in South America some half a million people suffered and several thousand died.

Poor sewage-disposal systems are nearly always to blame for this. A London lad whose feces contaminated a Broad Street well in 1854 when sewage somehow backed up into the water-supply system was demonstrably responsible for the tragic epidemic of cholera that followed. Around the same period, this disease was rife in many parts of the United States, while as recently as 1900, one person in five thousand died of cholera every year in the United Kingdom. The untimely death in 1861 of Prince Albert, consort of Queen Victoria, was (questionably) attributed to typhoid fever, consequent on the primitive sewage-disposal system at Windsor Castle, which in 1844 boasted fifty-three overflowing cesspits. (This situation, I can attest, has been considerably improved in recent years—as have now the fire-protection systems there.) A few years later, the Earl of Chesterfield also succumbed to typhoid fever, contracted while he was staying at Londesborough Lodge near Scarborough. Humbler souls tended to fare even worse. Around the same time, when the Brontë sisters were growing up in the little town of Haworth in Yorkshire, some 40 percent of their schoolmates died before the age of six; the high mortality—typical of the period—was doubtless correlated with the unsanitary arrangements in or around their houses. In the years following 1845, when many poor people emigrated from Ireland to the United States to escape starvation that faced them as a result of the potato famine, tens of thousands of them succumbed to cholera and dysentery in the American resettlement camps. In 1912 Wilbur Wright, having survived the perils of pioneering flight, succumbed to a diarrhetic fever, possibly typhoid.

In Holland, in 1671, Antoni van Leeuwenhoek, a former shopkeeper and civil servant who constructed the first simple microscope, reputedly examined a droplet of his feces during an attack of diarrhea. He thus became the first to see, and to illustrate and describe, the protozoan *Giardia*. Not infrequently, even today, infestations by this microscopic parasite (which produces relatively resistant cysts) plague American towns that have imperfect or damaged sewage systems, causing diarrhea among rich skiers in Colorado resorts no less than poorer folks at lower altitudes. In the winter of 1974–1975, several thousand people in Rome, New York, were afflicted in this way.

Regrettably, some cases of infection have been deliberately caused, as reported in 1997 by Török, Kolavic, and their respective coauthors. In 1984, religious activists laced salads at ten restaurants in The Dalles, Oregon, with *Salmonella;* seventy-five people got diarrhea; all eventually recovered. In 1996, someone used a culture of *Shigella* to contaminate muffins or doughnuts in a staff room of a Texas medical center, resulting in forty-five cases of diarrhea; here, too, fortunately none proved fatal.

Belladonna (which contains atropine) is among the most effective drugs in stemming the tide of diarrhea, though not of curing the disease that causes it. (But use this drug under advisement! An overdose of a few milligrams may result in a temporary numbing not only of one's bowels but even of one's whole consciousness, as I learned from experience some years ago when I fainted, for a few moments, in Thailand.) Tending to produce the opposite effect, perhaps, are the sulfurous waters of certain hot springs, renowned as cures for constipation. It has been asserted that the reputation of Karlsbad, in Germany, as a spa was built on the relieving of "undelivered feces."

Workers in Project Wildlife, a group of volunteers in the United States who look after abandoned or wounded creatures of various sorts, have noted that sick or starved birds often produce green droppings, colored apparently by biliverdin (a breakdown product of the hemoglobin in blood that occurs in bile, hence its name). They are often smellier than normal, a further indication of ill health. The droppings of stags before rut tend to be much sloppier in consistency than they are at other seasons. By examining such droppings, hunters can often learn not only about the species of animal they are hunting, but also about its well-being. A classic wildlife film made many years ago in the Belgian Congo showed a giraffe wounded by spears and excreting blood in its droppings; shrewd hunters, following it on foot, could thereby judge how close they were to their quarry and how weak it may have been when it defecated.

The nature of human feces is often a useful pointer to conditions of health or disease. The presence of "occult blood" in fecal smears, generally revealed by the presence of an enzyme that gives a characteristic blue coloration when mixed with a drop of benzidine solution, is often a sign of a bleeding cancer or ulcer in the digestive system. One of the personal physicians of King George III of England made a special point of looking for such symptoms, and noted a mauve coloration in the royal urine and feces, later interpreted as indications of a debilitating disease, hereditary porphyria. "Bloody fluxes" must have been much commoner among Europeans in earlier times than they are today. The number of putative remedies for such ailments in Nicholas Culpeper's *Herball* rivals those that alleviate a costive condition. (My copy of this classic work, a "new and elegant improvement" by a Dr. Parkins, was printed in 1810, but many of the recommended remedies must date back to the times of Paracelsus, if not earlier.)

Floating feces, although they are mostly just buoyed up by bubbles of colonic gases, are sometimes diagnostic of sprue or other diseases in which individuals deficient in pancreatic enzymes are unable to digest the fats and oils in their diets, which therefore tend to pass through unchanged.

Houseflies and bluebottles are usually to blame for the transport of pathogenic bacteria from exposed feces to human victims, usually by contaminating their food. It is not only human excreta that can act as sources of infection for mankind, and not only bacteria that are transmitted in this way. Well over one hundred different kinds of worms infest man and his domestic animals, causing illnesses of various degrees of severity. Most such worms are recycled through feces. In a monograph liberally illustrated with colored photographs, Thienpont et al. (1979) describe methods for the separation, concentration, and identification of eggs and other evidence of worm infections from the droppings of sheep and cattle, horses and pigs, dogs and cats, rodents and birds as well as man. (A medical friend of mine, whose job it was to scan fecal smears from American soldiers in the Second World War, found what he thought might be eggs of a wholly new kind of parasitic worm. One of his colleagues, with a little more botanical experience, was soon able to correct him: they were stone cells from pears.) Parasites claimed to be of an unknown species were found in fecal samples, reputed to have been produced by the fabled yeti, brought back from the Himalayas by members of the Slick-Johnson expedition.

Another colleague, supported by a liberal travel grant from the National Institutes of Health, once made an extended tour of South America, collecting samples of human excrements, which he brought back to his home laboratory for further study. He was investigating the distribution of antibiotic-resistant strains of coliform bacteria in remote areas where people had not been treated by modern medicines. Probably the majority of the cultures that he eventually isolated turned out to be coliform bacteria like *E. coli* and species of *Salmonella*. Most other kinds of human fecal microbes die within an hour or two after exposure to air, and so would not have given rise to colonies on his nutrient plates. I do not know what logistic difficulties he may have encountered when returning with his samples to the United States. I do know that a young lady studying coprophilous fungi at the University of Chicago some years ago was met with a degree of incredulity when she returned from Europe, where she had been collecting rabbit droppings, and exhibited her collection of treasures to the Customs officials in New York. (Many ecologists become accustomed to being regarded as eccentric.) I believe she finally convinced the authorities to allow them in, and in due course used information obtained largely from this collection to achieve an academic degree. Likewise, an eminent zoologist's duffel bag full of samples of jaguar feces raised a few eyebrows when it was being brought into the United States from Central America.

It is inevitable that, from time to time, things get lost. A batch of stool samples collected in paint cans, and being brought to Rockefeller University for analysis, was somehow forgotten and left in the subway. (History does not record whether they were used for painting graffiti, for which New York subway trains are now famous.) As for sending things by post, one is instructed to label such samples (somewhat gratuitously perhaps) with the declaration that they are "of no commercial value." Usually, this is undisputed. But when a scientist studying the coprophilous fungus *Pilobolus,* who had received samples of various mammalian droppings from a lady in Kenya, declined to pay even their shipping costs, she indignantly asked for them to be returned. And they were.

"Shit as a source of infection with an array of pathogens is too important to be dismissed because of its aesthetic failings"; so wrote Desowitz in 1981. In wild animals, parasites abound, and many of them end up in their feces. The pioneer microbiologist Antoni van Leeuwenhoek, peering through his lens at some infested frog droppings, noted that they were so full of living creatures of various sorts and sizes that the whole pellets seemed to move. The recycling of parasites and other pathogens through feces may create a serious hazard not only to the animals concerned but also to ourselves; it tends to be exacerbated when fresh human night soil is used as a fertilizer. As mentioned earlier, the incidence of hookworm infestation (caused by *Bilharzia,* a.k.a. *Schistosoma*) in rural regions of China was enormously reduced when the government insisted on the use of toilets for defecation instead of open fields and ditches (Morris, 1995).

A source of serious concern is the custom of adding buffalo dung to fishponds in Thailand to enhance the growth of the fish, notably tilapia, for eventual human consumption. Such farming methods tend to facilitate the recycling of parasitic bacteria, protozoa, and pinworms, tapeworms, roundworms, hookworms, etc. that depend on fecal transfer for their direct or indirect transmission, as spores or eggs, to new hosts. Migrating ducks have been shown to spread various strains of the influenza virus around the world, and it has been suggested that in mixtures of duck and pig excreta, such as one might find in Chinese farm ponds, there could conceivably be genetic recombination among viruses, leading perhaps to new and more virulent strains of influenza.

The phenomenon of "yellow rain," which gained notoriety at the end of the Vietnam War, eventually turned out to be a harmless mass defecation of tropical bees, sometimes induced when the weather becomes too hot and humid for their comfort. Each bee may lose as much as one fifth of its weight in the process. Flies, as housewives well know, excrete on mirrors and windows as well as in other places where the results are not so obvious. (However, the term "flyblown" should be reserved for meats and other materials that have received insect eggs, not their excreta.) Housefly feces are relatively harmless, but not all invertebrate excreta are benign. The frass of mites (not strictly insects, but close enough) constitutes one of the main causes of allergy in household dust. And that of a bloodsucking triatomid, the so-called kissing bug, may contain the trypanosomes which, when they enter the human body (through abrasions or scratches in the skin), cause Chagas' disease, a South American form of sleeping sickness.

Some recent investigations of human feces have centered around the drug trade, since a number of criminals try to smuggle illegal narcotics such as heroin in plastic capsules that have to be swallowed and later recovered. A hundred or more may be carried by one individual in this way, at some peril. If a single capsule ruptures, a lethal dose could be released; in recent years, there have been several deaths of would-be smugglers resulting from the rupture of opium-filled condoms in the wrong place. Even if this fate is avoided, convicted smugglers may end up in a penal purgatory.

15

USES FOR CONSTRUCTION AND DECORATION

The termites that infest trees and the wooden structures in our houses, and other insects like deathwatch beetles and their kin, usually kick out their frass, partly, one supposes, to keep their channels free for ventilation. However, ground termites also use their fecal matter along with soil and crumbs of wood to construct their nests, as do certain other invertebrates, like the exotic arthropod *Peripatus.* Many larval dung beetles, securely ensconced underground in their spherical cradles, use their own feces to line the growing hollow in the animal dung on which they feed, thereby retaining the integrity of their nursery walls. The South American ant *Dolichoderes* uses bits of other animal feces for a similar purpose.

Sometimes feces are used more or less incidentally for the reinforcement of birds' nests, such as those of cormorants in East African mangrove swamps. African zebra finches, however, use their droppings in a much more deliberate fashion to line their nests, and can make neat, pearly walls in artificial nest boxes. White-capped noddy terns glue their flimsy twiggy nests to tree branches with this kind of material. Secretary birds, which live in regions where soft plant materials like mosses are generally in short supply, often line their nests with the dried dung of large herbivorous mammals, such as buffalo and zebra. Other birds, such as gannets, kittiwakes, and South American oilbirds, make their nests almost entirely of their own droppings. Hornbills, which wall up their brooding females in hollow trees, generally use mud for this purpose, but when the wall is broken before the nestlings are fledged, the mother bird may use their feces to repair the damage. The young of the African yellow-billed species of hornbill do not merely provide the material, but also soon learn to collaborate in its utilization.

In the San Diego Zoo, small Australian marsupials called "feather gliders" have been observed to use their fecal pellets, which are attached to the glass sides of their enclosures, as stepping stones, though I doubt whether they have much opportunity to do this sort of thing in nature as they leap around among the eucalyptus flowers.

In salt-evaporation ponds, the deposition of fecal matter by fairy shrimp (*Artemia*) helps to stabilize the bottoms of the pools by lining them with a relatively impermeable material. Without it, much of the brine would soak into the ground and be lost. Humans of many races, notably the

Masai, Dinka, and Nuer tribes of Africa, use cattle dung as a mortarlike binding agent for their round wattle huts as well as for fuel. Within living memory, the thatch on cottage roofs in Byelorussia (now Belarus) was compacted with cow dung, while, even nearer home, the same material was used by the eminent ornithologist Alexander Skutch to plaster the walls of the home he built in San Isidro de General, in the hills of Costa Rica. Probably because it contains residual hemicelluloses along with gut bacteria, dung constitutes a valuable addition to the mud used for flooring in much of rural India. It serves as a compacting agent, and when properly applied, it renders floors smooth, hard, and durable, and considerably reduces the dust that would arise if mud alone were used. The Nuer even use dung to coat their cheese-fermentation vats, reportedly to keep out flies, though one may question its effectiveness for this purpose.

Other uses of cow dung are legion. Among the Masai, a crown of dung may be worn by a bride, while for everyday use a poultice of cattle dung is used to stanch the bleeding of oxen after a neck vein has been "tapped" to provide liquid sustenance for their owners. Cow dung also constitutes the basic material for some sculptures recently on exhibition in Cardiff. In rural areas of Wales, it is often abundant; it is plastic when fresh but hardens when dried, and is relatively inexpensive—so why not? Elephant droppings, although less readily available than cow manure in Britain, have been successfully used by a painter who in 1998 was awarded the Turner

Prize; whether for the originality of his imagery or his choice of medium is anyone's guess.

An infusion of donkey droppings was used by Michelangelo to give the appearance of antiquity to some of his marble carvings, a technique still applied today by less reputable artists involved in faking antiques.

Some rotifers and caddis-fly larvae neatly incorporate their droppings into cases, thereby using their waste materials not only for protection but also for camouflage. Bagworms, which are the larvae of certain moths quite unrelated to the aquatic caddises, do the same sort of thing on dry land. A predator of termites with the awesome name "assassin bug" covers its back with termite droppings, presumably so that its prey will be lulled into thinking that it is merely a midden of their frass. Some beetles, after laying a batch of eggs on their food plants, cover them with feces, presumably to conceal them from any egg-eating predators. The larvae of other beetles, species of *Cassida,* carry over their backs little parasols made of their fecal pellets—whether for shade or camouflage, or both, has not been established. The caterpillars of hairstreak butterflies, as their last act before pupating, cover their chrysalids with frass, supposedly with similar objectives.

16

USES AS FUELS, ETC.

In many parts of the world, dry manure is burned as fuel, since much of its organic content, although indigestible, is still combustible. Indeed, animal droppings are a useful commodity in many societies. The Abyssinian maid who in the well-known poem by Samuel Taylor Coleridge, "Kubla Khan," played on a dulcimer, must have been of a privileged class: a more usual occupation of such maids today is to sweep up goat droppings, to be used alone, or blended with straw, as fuel. (Coleridge, I believe, never visited Abyssinia. The nearest he got to that country was Malta, where today the abundant feral goat droppings in the Maltese countryside go mostly unswept, now that other domestic fuels are more readily available.) Without animal dung as fuel, many of the peoples who live on dry steppe lands and pampas, above or beyond the tree lines in Central Asia and

South America, could not long survive unless they were wealthy enough to pay for other sources of energy and fuel, such as electricity, coal, or oil. With no wood or peat to burn, Tibetans must collect enough dung to heat their yurts and cook their meals. Their lives are therefore absolutely dependent on browsing animals, which not only convert pasture grasses and forbs into protein for milk, meat, hides, and furs, but also considerately package the residual cellulose as combustible droppings.

Under the aegis of the god Lakshmi, cow dung is an important commodity in India, where it constitutes some 25 percent of the total domestic fuel used in rural areas, or the estimated equivalent to some thirty-five million tons of coal per annum. In many parts of India it is the sole source of combustible material, and it is listed among the five useful products of cows, along with milk, curds, butter, and urine. In some villages, women carry on their heads baskets containing cakes of this fuel, which they sell for a living. The cakes have been referred to as "bricks of gold." (Conversely, in ancient Mexico real gold was called "god shit," or native words to that effect. In Nahuatl it is still called *teocuitlatl,* which has the same meaning, while *xicohcuitlatl* is the word for beeswax.) The Hopi Indians, at least until recent years, used to buy from their Navajo neighbors sheep droppings for fueling their pottery kilns; the ancient Egyptians used a similar fuel in their lime kilns, too. Llama dung, called *tarquia* or *bosta,* is still used to fire the motors of a steamer plying the waters of Lake Titicaca in Peru. Cattle or llama dung has been much used as fuel for cooking; for aesthetic reasons, if for no others, it seems preferable to human feces. When God instructed Ezekiel (4:12–15) to bake his bread on a fire fueled by human feces, and the prophet demurred, he was assured that it would be permissible to use cattle droppings instead.

Yak dung is widely used as a fuel in Tibet, where it is collected (generally by women) and sometimes stored by the sackful. It is reputed to burn more readily than, say, the droppings of sheep or goats (which they call *rima*). Whether the smoke of burning yak turds adds a distinctive flavor to cooked sheep brains, as hickory smoke does to broiled steaks, is a subject on which some Tibetans may have strong views. *Kashanjiagan,* buttered crepes toasted over a horse-dung fire, gain a flavor especially appreciated in Kirghiz. As recently as the American Revolution, and well into the latter part of the nineteenth century, many of the farmers in the Great Plains depended on cow dung as fuel for cooking and for heating their homes, just as the native Indians had used bison droppings, or "buffalo chips," in those same territories before their lands were expropriated. At the time of the Mexican wars, Cortés reported seeing human excreta for sale, supposedly to be used as fuel, in canoes along the canals of Tenochtitlán. Even the ash of dung has been found useful, for instance, as a source of salt and as a dentifrice among the Latokas and other tribes along the White Nile in Africa.

Perhaps significantly on the National Day of the People's Republic of China, in 1991, an explosion at a public toilet (women's section) on the outskirts of Beijing was attributed to methane accumulation and, presumably, a carelessly lighted match. We have already mentioned that Thomas Crapper once suffered a concussion in similar circumstances. (Methane is the gas referred to earlier in the section on gaseous components, only in these cases it was produced outside, not inside, the body.) Collected in various ways, including "composting latrines," it has been—and in some places still is—extensively employed in agriculture and aquaculture, and for the production of biogas as fuel.

Methane, or biogas, produced in sewage fermenters is a valuable source of energy, since the gas can be burned and used to cook food (as on many farms in India, Malaysia, and China) or, on a much larger scale, to produce electricity (as for example in Denmark, Los Angeles, and in Cook County, Illinois). For this purpose, a liquid preparation sold in Hong Kong is supposed to promote the rate of degradation of pig droppings, clearly an invention for which the world has long been waiting. Various possibilities for generating energy from farm animal manure are discussed in a 380-page report prepared a few years ago by the NEOS Corporation for the U.S. Department of Energy. Gasification and anaerobic fermentation systems are compared and economic analyses are presented, together with a useful bibliography and quantitative data on the farm fecal productivity of every county in thirteen western states. In the 1960s, an Englishman in the West Country achieved some notoriety by marketing a device which, fitted to the roof of a car and periodically filled with chicken droppings, produced enough methane to fuel the motor—or so he claimed. A similar device, apparently a modified carburetor, was recently on sale in Oregon for as little as $33. Along the same lines, the Rocket Research Corporation of Los Angeles studied the possibilities of using fecal-fermentation methane as a propellant for space travel, though it apparently dropped the scheme eventually.

On a small island off the Norwegian coast, the author Roald Dahl observed (and smelled) a man smoking shredded goat dung. The incorporation of dried animal droppings as an adulterant in tobacco started in the New World, but along with the use of the weed, it spread to other parts of the world. Similarly, dried poultry droppings have been used to adulterate opium. Joseph Needham (1978) recounts that certain medieval Chinese "bombs" consisted mostly of finely powdered human excrement mixed with castor oil and arsenic, a concoction which, though not of itself explosive, could have been combined with gunpowder for effects that might eventually be disruptive of human systems. On the other side of the coin, Robert Oppenheimer is said to have told Leo Szilard, "The atomic bomb is shit." (These items can hardly be regarded as directly relating to fuels, but I don't know where else to include them in this book.)

17

USES AS FERTILIZERS

Perhaps the best-known use of feces, both human and other animal varieties, is as fertilizers, which have been widely employed as an important part of farming practice since time immemorial. In March 1996, symposium workshops were held under university auspices in Wetumka and Woodward, Oklahoma, specifically to discuss pig manure and its characteristic odors. Camel dung has also served as a source of ammonia, well before we learned how to make ammoniacal fertilizers commercially out of thin air.

Human fecal material has been euphemistically called "night soil," though produced at any time of day, and its collectors—at least in England—were called "night men" until their recent elimination by what amounts to technological unemployment. It was generally collected under cover of darkness, sometimes as ordained by local laws (such as those enforced until relatively recently in London and Paris). Along with horse and cattle dung it is extensively used as a fertilizer, especially in China where, under a Maoist decree, it was to be regarded as property of the commune and not just of the individual producers. As I can attest from personal experience, a pair of sewage-filled buckets, carried on a yoke over the shoulders, is no mean burden. (I didn't take it far. My Chinese hosts were afraid I might spill some of the valuable contents.) It is smelly, of course, but during the so-called Cultural Revolution its handlers were assured that, if they kept the edicts of the Great Leader in mind, they wouldn't notice the odor. (I couldn't confirm this.) For further details on public health and agricultural campaigns based on "the excrement of the people of China," the reader is referred to a fascinating contribution to unconventional history by Morris (1995). There he tells of the five hundred million tons of manure that the Hunanese managed to collect in a mere ten weeks, and of the valiant night-soil carrier Shi Chuanxiang, who rose thereby to become a member of the Beijing City Committee of the Chinese People's Political Consultative Conference, and to be photographed shaking hands (washed, one hopes) with the Chairman himself.

The Bhanji (removers of ordure) are, or were, among the Untouchables for whom Mahatma Gandhi understandably expressed considerable concern and sympathy. They are still not highly regarded, however. My late colleague Professor John Isaacs asserted that "the scientist is being forced into the position of the man with the bucket and shovel behind society."

In other parts of the world, too, night soil has often been put to good use. Around Marseilles in the early eighteenth century, the cess of galley slaves was collected and used as fertilizer on the land, as it was considered too valuable to be allowed to fall overboard. In the musical *Miss Saigon,* a leading character sings, "You can sell shit and get thanks," although regrettably, for our purpose, he gives no indication of its provenance or price. Indeed, Leonardo da Vinci once pointed out, cynically but probably correctly, that for most people their only useful contribution to society is to their local cesspit.

Sir John Russell, when director of the Agricultural Experimental Station at Rothamsted during the Second World War, estimated that the flushing away of human feces wasted seven shillings and ninepence worth of potentially rich fertilizer per capita (so to speak) per annum. With a yearly production of approximately thirty million tons (a reasonable estimate for the United Kingdom), its worth is not to be sneezed at. A dried and relatively deodorized product of this sort is marketed in North America as "Milorganite," and doubtless similar preparations are sold elsewhere under other names. For instance, although the treated waste water of New York City is discharged into the Atlantic Ocean, since July 1998, 100 percent of the biosolids produced have been recycled, some of it dried and trucked to rural sites, for example, Sierra Bianca in Texas, despite complaints by residents with more olfactory or aesthetic sensitivities. (But if it improves the crops, so what?) For inland cities, of course, discharging municipal wastes into the sea is not an option, and in the United States one has to resort to sewage treatment plants.

In Roman times, Pliny recorded how much the farmers valued *stercus* from cowsheds, which they spread on their fields and thereby improved their crops, though I doubt whether it was really King Augeas of Greece (who reputedly kept three thousand head of cattle) who first promoted this practice. In a biblical reference (Luke 13:8), we are told of the beneficial effects of dunging on fig trees; in Peru, human excrement is supposed to be especially good for the growth of maize. And as Shakespeare wrote (*Timon of Athens,* Act 4, Scene 3, lines 446–48): "The earth . . . feeds and breeds by a composture stolen from general excrement." According to Henry Ford, not usually regarded as an authority on agriculture, "You raise your own fertilizer." Or, if you don't, you can usually find a source from which you can buy it. Our local San Diego newspapers often carry small advertisements for horse manure, free for the taking.

Mushroom growers are well aware of the value of horse dung as a substrate for their fungus cultures: one wonders what will happen when the supply of this valuable substrate has been completely replaced by automobile emissions and worn-out car tires. (Incidentally, we are not the only animals to cultivate edible fungi on feces. Leaf-cutter ants do so, too, carefully tending the crops that they raise on chewed vegetation and their own droppings in the cellars of their nests. When a young queen goes off to start a new colony, she takes bits of fungus along with some of her own feces to keep it growing, and on this lays eggs from which her workers will ultimately emerge.)

Until recently, in the countryside of the Netherlands, where the number of pigs and cows approaches that of Dutchmen, one could see specially constructed machines spraying liquefied pig manure over fields that would later blaze with the colors of well-fertilized tulips. There, however, the supply considerably exceeds the demand. One is then faced with the problem of disposing of the excess manure. Other ways have to be found for dealing with the stuff. There seems to be too little arable land for the disposal of all the fecal matter from five million Dutch pigs and twelve million head of cattle, and it has to be trucked away somewhere. Permits are needed for its disposal. It has been reported that since 1987 the problem has become so acute that farmers have been known to include, among the desirable attributes of prospective spouses, the possession of requisite permits for pig-manure disposal. Pig-manure spraying is no longer legal in the Netherlands, although it is still practiced by some farmers after dark. The liquid must now be injected into the ground, where its efficacy as a fertilizer is unimpaired. Fertilizing the North Sea, another current expedient, may be a good thing for the local fish; but, when there is a brisk westerly wind, people on the southeast coasts of England are less than enthusiastic about the whole scheme.

In Canada a liquid effluent of swine manure has been found to be particularly beneficial for the continuous culture of microscopic algae. A firm in the United States, Farm Builders, used to collect and distribute the liquid effluvia from piles of farmyard manure—it was euphemistically called "serum"—for the cultivation of algae later used as soil conditioners. In England a relatively deodorized product is sold under the name "Cowpact," marketed under the unambiguous motto Direct from the Cow; while in Singapore one can buy plastic bags full of a rich black compost called "zoo-poo," the source of which is not hard to guess. Cow and chicken manure, as well as a mixture of assorted dungs from the San Diego Zoo, is marketed by Best Soils of Chula Vista at $17 per cubic yard; among other benefits, it saves the City of San Diego some $15,000 annually in landfill costs. A farmer in Inner Mongolia told me that he sells sheep droppings, for the American equivalent of about five cents per kilogram, to a trader who trucks them south and resells them as fertilizer at a 300 percent markup. I recently toured a large farm complex in Brittany that specializes in a variety of blends of pig manure with seaweeds; suitably mixed, matured, dried, bagged, and sold as fertilizer, they bring in an appreciable annual income.

The fertility of England's famous chalk downs, like those in the Cotswolds, is said to be considerably improved by the droppings of the sheep folded on them, though it is not clear how the net content of nitrogen and phosphorus could be increased by this sort of recycling. Incidentally, some Englishmen seem to have considerable faith in the efficacy of cow manure. To remedy the relatively low stature of their church spire, the villagers of Coombe, in Oxfordshire, were reputed to have piled such fertilizer around its base in the hope that the spire would be stimulated to grow. (This, at any rate, is a story neighboring villagers recount to exemplify the gullibility of Coombesmen.)

In some cave systems, deposits of bat guano (chiropterite) may exceed 10 meters in thickness and have the potential of yielding hundreds of thousands of tons of excellent phosphatic fertilizer if excavation and transportation problems can be solved, and provided one takes cognizance of the possibilities of its being contaminated with rabies virus. After the discovery of the enormous Lechuguilla cave complex in New Mexico in 1914, large quantities of bat manure were mined from its capacious antechambers. Earlier, during the American Civil War, this material was used, after being calcined with potash, as a source of saltpeter for the manufacture of gunpowder, and in some karstic areas in China it is still used for this purpose. In certain Peruvian caves, artifacts of the Mochica Indians have been found embedded in bat-guano deposits at depths of 19 meters. In some places, the deposits may be two or three times as thick as this, and the rates of accumulation could exceed a couple of centimeters per year. Even bigger piles have been produced by horseshoe bats in Southeast Asia. And at least one monastery in Thailand supplements its income by selling bat guano from nearby caves, thereby earning the equivalent of many thousands of U.S. dollars annually.

Bird droppings are of course a natural source of fertilizer for the land or ponds below. Persians have long used pigeon dung as fertilizer in their melon fields. In some desert areas, the droppings of owls and other passing birds are the only sources of nitrogenous matter that they ever receive. Since the protein-rich diets of seabirds, which feed mainly on fish, generally contain more nitrogen and phosphorus than the birds need, excess amounts of these elements have to be excreted. Enormous quantities of good fertilizer are deposited by goony birds or boobies, gannets, pelicans, gulls, and terns along the dry coasts of South America and South Africa, and on a number of oceanic islands (notably Nauru, Christmas Island, and the Pribilofs). Some of the few human residents of Midway Island have reported that their life there is rather like living in a birdcage, "except that nobody changes the sand or paper."

Guano is the accumulated droppings of millions of seabirds that throng many rocky shores. Although some 95 percent of it falls directly into the ocean, where the fertilized waters become rich in phytoplankton, zooplankton, and ultimately the little fish that serve as food base for the birds, much of it still accumulates on land, where rainfall is insufficient to wash it away. In some places, the guano deposits may be more than 200 meters thick! Correspondingly, some areas of the sea bottom off the Peruvian coast are covered with a thick sludge of fish feces, mostly from the anchovies that have survived predation by birds and seals. For centuries the mining and exportation of guano (a word of Quechua Indian origin) has been one of the main sources of income in Peru. Ecologist G. Evelyn

Hutchinson compiled a 554-page illustrated thesis on "The Biogeochemistry of Vertebrate Excretion" (1950) in which he hardly even scratched the surface of this vast subject. Since about 1840 the guano trade has had considerable impact on the economy of Peru (extensively reviewed by Gootenberg, 1989); it still exceeds 100,000 tons per year.

When visiting large modern poultry farms, one is often impressed by the stately snow-capped sierras of waist-high chickenshit accumulating under the seemingly endless rows of caged hens. Every year some 300,000 tons of chicken manure is produced in Arkansas alone, and at today's U.S. prices it can fetch as much as $70 per ton. Poultry farmers in Hong Kong sell some of this material to worm farmers, who feed it to the bloodworm cultures that ultimately provide wriggling food for aquarium fish. Although chicken droppings from large poultry farms are ordinarily used as fertilizer (and sometimes as a major supplement to the diet of cattle and sheep), farmers have to bear in mind that the higher its content of nitrogen, the less efficient must be the birds' metabolic conversion of dietary protein to egg protein.

In the waters of Long Island Sound, duck excrements are among the major sources of fertilization or pollution, depending on whether one approves or disapproves of the dense growth of phytoplankton that often develops in the Sound during the summer months. In the Far East, in mixed farms duck droppings are generally among the principal fertilizers of fishponds, where they encourage the growth of algae and small invertebrates that constitute the main foods in the aquatic part of the farm system.

If we were to argue teleologically, we could say that many kinds of trees, conscious of the great value of animal droppings as fertilizers, produce large, succulent fruits, rich in carbohydrates (which are metabolically "cheap" to synthesize) though poor in nitrogenous components such as proteins, to attract birds, fruit bats, and other animals. These creatures tend to sit around on the branches, gorging themselves on the succulent fruits while depositing a liberal repayment of nitrogen-rich droppings around the trees' roots. Special "traps" for insect droppings are formed in nature on the branches of certain New Guinea lianas, which produce special urn-shaped "toilets" for ants. The debris that accumulates in them forms a rich compost into which grow adventitious rootlets of the vine—a symbiotic sanitary arrangement from which both parties profit.

Earthworms are of course among the major overturners of soils and compost in many temperate regions of the world. In the Netherlands, worm casts from specially accumulated piles of garbage are sold commercially for fertilizing gardens. At worm farms in South America, one can buy not only earthworms (fish bait for anglers) but also bags of their casts (fertilizer for potted plants); the latter may fetch as much as U.S. $1 per kilogram. In western Australia, I have seen elegant bottles labeled "worm wee," the liquid effluent of worm farms, sold for the same purpose. However, where I live in California, most of the leaves in our compost heap, being relatively dry, are converted into frass by sow bugs (isopods); worms are relative rarities. Wherever silkworm caterpillars are cultivated, their droppings make a valuable fertilizer. I have heard that in the markets of old Delhi, in India, there are stalls that specialize in selling insect droppings, but I have been unable to establish whether these are silkworm feces. Maybe they also sell frass of stick insects and locusts, which are among the myriad natural products used in Chinese traditional medicine.

In the Negev Desert of Israel, where most other animal droppings are in short supply, the feces of snails that graze on the thin layer of lichens on rocks have been found to play an important ecological role, daily adding as much as a milligram of nitrogenous fertilizer per square meter to the sandy soil.

18

PUBLIC NUISANCES

Unlike cats, dogs do not habitually bury their feces, perhaps because wild canines range widely, whereas felines, which tend to live in lairs, prefer to keep their precincts reasonably clean and apparently try to conceal their whereabouts from other animals. In a few cities, notably Reykjavík in Iceland and Beijing in China, the keeping of dogs is forbidden except in special circumstances, and as a consequence the streets are comparatively clean. But in most other urban areas domestic dogs can be real nuisances. I have noted that the pavements (sidewalks) of London can get pretty messy, although few are as sullied as those of the wealthier (and therefore doggier) suburbs of Amsterdam or Liège in Europe, Buenos Aires in Argentina, or Dee Why in Australia, where dog-dropping densities may exceed 0.3 per pavement meter. A Chinese scientist recently returned from a three-month visit to Belgium, and when asked for her reactions to that country, she mentioned in her first sentence the deplorable abundance of dog feces that she had observed on the city pavements.

When a contemporary British poet, John Sparrow, wrote in a letter to *The Times* (September 30, 1975) about "that indefatigable and unsavoury engine of pollution, the dog," we did not need to be told into what he had recently trodden. My favorite poetess, Dorothy Parker, as I learned from a television biography, failed to housebreak her pet dogs, whereupon she went down a couple of notches in my esteem. Actually, dog droppings are not just aesthetically offensive, they also present hazards to human health (Marron and Senn, 1974). They may contain cysts of the protozoan parasite *Giardia,* which could be washed into drains and watercourses and ultimately find their way into sources of potable water. (As mentioned in Chapter 14, giardiosis is one of the causes of diarrhea in humans.)

Many cities nowadays require that dog owners, or more specifically dog walkers, clean up their messes as soon as their pets have finished generating them. I was told of a lady in our town who takes daily walks with a large dog, a small dog, a shovel, a large plastic bag, and a small plastic bag, but I am not sure that such specialization is warranted. Efforts to educate the pets' owners are sometimes almost as ineffective as trying to educate the dogs themselves. A recent study by social psychologists at DePaul University, in Chicago, indicated that of the twenty pounds of dog shit deposited weekly on one section of a city block, only about 5 percent was picked up by the dogs' owners, leaving city sanitation teams, aided by wind and weather, the passage of time, and the shoes of unobservant pedestrians, to dispose of the rest in due course. (And we call ours a civilized society?)

Many of the street cleaners of Paris are equipped with special "pooper scoopers" for clearing pavements of canine filth. Some city parks elsewhere have special sandpits designed to be used as dog toilets, although it must be admitted that many dog owners, and presumably all their pets, fail to read and observe the signs. (For a review of further aspects of this problem, see Jason et al., 1979.) In Monaco some of the public facilities for canine pets are quite elegant.

Dog diapers, while rarely employed for various reasons, are not unknown. To keep dogs out of private gardens, some people, for obscure reasons, recommend displaying a bottle of water, while a recent invention from Queensland involves the use of ultrasound as a repellent. The pet dog problem blights not only streets and gardens. The oceanographer John Isaacs observed that whereas "the grey whale leaves tons in its wake, most of the fecal matter on California beaches is put there by dogs."

Herons of various kinds have earned the rural appellation "shite-pokes" (cf. the Pennsylvania-Dutch word *scheit pock*), apparently because of their liberal production of excreta. When liberated in flight they may form chalky lines on the ground. The droppings of herons can accumulate to such an extent that they eventually kill the vegetation underneath their nests. Fish-eating birds, many of which nest on sea cliffs, excrete over the edges of their perches, thereby whitening the rocks below, as do town pigeons, whose ancestors were coastal cliff dwellers. (A special paint preparation has been developed recently in Sweden which, when applied to walls, renders them more easily washed free of bird droppings as well as graffiti.) For millennia, town pigeons have plastered snowy caps on statues, temple ledges, church cornices, and bank arches, where their accumulated droppings (called "mutes" by the cognoscenti) may accelerate erosion. On monuments they are an unmitigated nuisance. On Nelson's Column, in pigeon-infested Trafalgar Square, the removal of half a ton of droppings costs the city of London about £35,000 every year. (I developed a special unfondness for such mutes when, as an assistant in the Wood Pigeon Investigation in Oxford, I had the job of cleaning out the cages on the museum roof.)

Pigeon droppings sometimes contain small pebbles that have passed through the gut from the gizzard, where they normally serve as grindstones for seeds and other hard elements in the diet. The presence of grit further exacerbates problems of dispersal and removal. Pigeon debris has even been reputed to cause fires among accumulations of nesting rubbish in attics. This was noted in the second century by the Greek physician and writer Galen, and more recently, in 1595, was supposed to have caused the conflagration that burned the cathedral of Pisa. In All Saints Church, in Mattersey, Nottinghamshire, some of the vents near the vaulted ceiling admit not only fresh air but also resident bats, which, unwilling to confine their flights to and from the traditional belfry, tend to distribute offerings on the altar and elsewhere. Under a recent British law, bats may not be constrained; therefore, conflicting interests of congregation and conservation are apparently at an impasse.

Once, a century or more ago, the increase of horse-drawn vehicles on city roads led doomsayers to predict that traffic circulation in civilized centers might ultimately be brought to a standstill by the accumulation of horse droppings. Fortunately, perhaps, the invention of the internal-combustion engine resolved that problem for us, although it did leave civilization with another pernicious health hazard, the gaseous excretions of automobiles. Nowadays, in China, although much produce is carried around in horse-drawn carts, horse droppings are rarely in evidence on city streets, since local legislation, and the threat of stiff fines, require that most such beasts of burden be equipped with rear-end equivalents of nosebags. Similar strictures apply to the carriages that take tourists around Prague. Indeed, in many cities horse droppings have become rarities. I have heard of a resident of Westlake Hills, Texas, who formally petitioned his city council to allow him to keep a pet donkey at his home so that his children would grow up knowing what manure looks like.

Bucolic sanitation is no recent problem, of course. If we are to believe the biblical account, sanitary conditions in the hold of Noah's ark (constructed, we may note, of shittimwood) must have become unbelievably noisome toward the end of its six-week voyage to Ararat. Elsewhere we are told that Augeas, an ancient Elisian king, allowed the dung of his three thousand oxen to accumulate for thirty years until an estimated 100,000 tons had raised the floor level of their stalls to a height of a meter or more. Heracles, invited to clean the place out, wisely chose not to rely solely on the use of shovels: instead, he diverted the flow of a local river, the Peneus (some say also the Alpheus, though that stream runs 25 kilometers south of Elis), which washed things out fairly completely within a day. This feat of sanitary engineering was commended by many, though perhaps not by those living downstream.

Travelers who fly into Cairns, in Queensland, are often left in no doubt as to why the small airfield is called Cowpat Strip, since its use is shared by cattle as well as planes. Indeed, some seasons and sites in Australia are renowned for their dung flies. In Los Angeles, flies are usually less pestilential than in Australia, but the mountains of dung produced in feedlots in that area seem to be growing daily, yearly, with no end in sight. It is apparently now more economical to use nitrates or urea to fertilize fields than to recycle this rich organic source of natural fertilizer. However, eventual uses for it may be found elsewhere. For instance, cow manure, dried and spread on shores that have been polluted with spilled oil or tar, could accelerate the dispersal of the oily matter and its degradation by bacteria and yeasts.

19

MYTHS, LEGENDS, AND HOLY ORDURES

Hippocrates asserted that pigeon droppings were efficacious against baldness, though whether this claim was based on personal experience is not recorded. Peacock feces were supposed by Pliny to cure such ills as fevers and epilepsy. He recommended hippopotamus droppings for the same purpose. Warm donkey droppings were prescribed in Egypt, at least until recently, to alleviate soreness of eyes. (Their efficacy would seem to be highly questionable: one might think such substances would have the reverse effect.) An aqueous mixture of horse dung with various astringent herbs is, or was, one of the Maya recipes for promoting the expulsion of afterbirths. Today in some countries, bird droppings, when they land on someone, are supposed to bring good luck, though I have not been able to confirm this from personal experience. Shortly after we were married, my wife and I were sitting under a shady tree in the Luxembourg Gardens of Paris when such an auspicious (?) event befell my wife. She urged that we move at once, but I saw no reason for doing so, since lightning is not supposed to strike twice in the same place. Five minutes later, I was proved wrong: my wife was blessed with a second "lucky" event, and I was never forgiven.

When the Indonesians recently built a seagoing prau designed to retrace prehistoric sailing expeditions between Sulawesi and Madagascar (8,000 kilometers—no mean feat in ancient times), they secreted into her keel some chicken droppings, along with gold and silver, to ensure the good fortune of the venture, which in fact did prove successful. Perhaps also to ensure good fortune, the kings of Nepal on inauguration were (and perhaps still are) anointed with horse and elephant dung, which may be no less effective than the holy water employed in other kingdoms.

At the annual fair in Llanigon, Wales, one of the traditional events is the "bull drop," in which wagers are laid on which area of a specially squared-off pen is first to receive a fecal deposit. In 1991, a similar event took place in Hendersonville, North Carolina, where the Rotary Club sold tickets, doubtless in the name of a good cause, for what they called a "cow-patty bingo." And in Siena, Tuscany, where before the annual horse race the beasts are led into a church to be blessed, it is regarded as a favorable omen for one of the horses to leave an offering during the service, although one doubts whether the verger who has to clean the stalls appreciates the good luck conferred thereby.

There is a story about a donkey that, led onstage perhaps to represent the rustic cavalry in a production of Mascagni's opera *Cavalleria Rusticana,* defecated on the proscenium. The eminent conductor, Sir Thomas Beecham, stopped the performance at this point by tapping on the rostrum; he then turned to the audience and announced, "Ladies and gentlemen, a moment's reverent silence, please. We are in the presence of a superior being: a critic."

One Kashmiri legend tells of a noble youth who daily excreted seven jewels (of an unspecified nature), while some Buddhists have claimed that the feces of the Lord Buddha glowed in the dark, doubtless as evidence of their intrinsic holiness. And in this connection we should also recall the reputed sweet smell of certain virginal Roman Catholic novitiates' excreta mentioned earlier. Stercoranists used to believe (perhaps some still do) that some of the consecrated elements of the Eucharist were so indigestible that they survived undegraded, like cellulose, as they passed through the human digestive system, though I know of no confirmatory evidence for this. (The expletive "Holy shit!," used for example by a witness of a particularly vicious mass murder in a New York train, is a not uncommon combination of blasphemy with coprolalia.) In *Pilgrim's Progress,* the pious John Bunyan refers to a purge made from the body and blood of Christ, but we must assume that this should be taken spiritually rather than literally.

Bourke (1968) has provided some interesting documentation about what he calls the "alvine egestae" of the Grand Lama of Tibet, dried samples of which were packaged in little boxes or in small bags suitable for being carried around one's neck. Some were also used as condiments or snuff. They were supposed to bring good luck, although a Dr. W. M. Mew of the U.S. Army, who analyzed a sample on the eighteenth of April, 1889, reported that he could find nothing specifically to account for this magical property. The feces of a chief patriarch of Araby, King Afrida, were supposed to be equally efficacious, though on equally dubious grounds.

Some ancient Egyptians (a couple of millennia before Harvey elucidated the circulation of the blood) had the rather odd idea that feces were formed in the vessels coming out of the heart. Excrements seem to have held a special fascination for the pre-Columbian Americans of Central America, where the Mexican goddess of love and dung, Tlazolteotl (counterpart to the Roman deity Stercutius), was revered and commemorated with elaborate illustrative designs. A treatise on the subject by A. Austin Lopez (1988) consists largely of a compilation of items of fecal information from the native lore of Mexico. They were translated from Nahuatl and Maya texts into Spanish, and illustrated by Francisco Toledo with an abundance of grotesque vignettes.

Defecation is generally regarded as a far from spiritual act, although orthodox Hindus say a special prayer before excreting; Jews after the act say one, the *asher yatser;* and Muslims (to be on the safe side, perhaps) say one at both times. In the early 1930s, when the Leahy brothers visited New Guinea, the natives originally considered them to be spirits from another world, and were disabused of this idea only when, after the Europeans had departed, their feces were dug up to reveal their human nature. The German writer Heinrich Böll wrote of a girls' boarding school in which Sister Rahel claimed to be able to diagnose even the scholastic performance of her pupils by examination of their feces. We may also note that Peruvian hill farmers have used the dispositions of sheep dung for copromancy, while in a fable of the Maidu Indians the coyote examined his own droppings for similar reasons.

ENVOI

I started this book by telling readers about the neat appearance of moose droppings, and how they set me considering the shapes and sizes of feces in general. This, in turn, led me to wonder about other features of fecal matters. Then it struck me: why is it that I, a biologist, have begun only so late in life to consider such topics? Why have others apparently given them little or no consideration at all? A notable exception is an exhibition of various fecal matters, called Poepgoed (pronounced in Dutch, and meaning in English, "poop-good"), originally compiled in Leeuwarden in 1996 and subsequently exhibited in museums at Harderwijk and then at Woerden, where it may still be. When you are in the Netherlands, try to pay it a visit.

Bookshelves are full of books on food and sex, and surely excretion is an equally essential part of our lives and of almost all other animals'.

So with this book I have taken it upon myself to try in small measure to remedy the deficiency.

On this vast subject perhaps one could arrange a conference, or issue a series of postage stamps, or dedicate a wing in the Museum of Modern Art, or form a society with a newsletter, or—is it too much to ask?—a journal: *The International Journal of Comparative Coprology.*

ACKNOWLEDGMENTS

I should like to acknowledge valuable suggestions in the preparation of this essay from many colleagues, including Drs. M. Baars, K. Benirschke, L. Cheng, M. Coe, A. Dundes, R. A. Farrand, H. Fernando, G. E. Fogg, J. H. Frank, J. Goering, R. N. Hamburger, W. M. Hamner, A. F. Hofmann, N. D. Holland, D. K. Jordan, V. Kytasty, L. D. Newmark, K. Norris, C. M. Perrins, W. H. Propp, A. L. Rice, D. Rimlinger, V. Roussis, P. T. Robinson, P. R. Saltman, J. Tiffany, B. Tyrrell, C. Wills, M. and D. Warrell, and other friends, including A. Engelman, A. Francis, L. Heinlein, D. Hester, S. Hinton, E. Jordan, M. S. Morris, D. H. Murphy, and W. Cheng Ward. Aaron Borovoy, at the University of California at San Diego, was indispensable in helping me to process this book through its early stages. My agent, Dr. David Madden, recognized its possibilities, and Robert Loomis expertly—albeit ruthlessly—edited it for publication.

BIBLIOGRAPHY

"It would be a precious exercise to provide specific documentation for every detail of this commentary; it would be both arduous and redundant—many single points would deserve a library" (Lederberg, 1993). However, some of the more useful references dealing directly or indirectly with the subject of coprology are listed below.

Albone, E. S. *Mammalian Semiochemistry.* Chichester, England; New York: Wiley, 1984.

Arango, A. *Dirty Words: Psychoanalytic Insights.* Northvale, NJ: Aronson Inc., 1989.

Austin Lopez, A. *Una Vieja Historia de la Mierda.* Mexico City: Ediciones Toledo, 1988.

Bang, P., and P. Dahlstrom. *Animal Tracks and Signs.* (Translated and adapted by G. Vevers). London: Collins, 1974.

Beard, D. B., and J. Gatts. "Effects real and relative of a space-type diet on the aerobic and anaerobic microflora of human feces." *Aerospace Medicine* 37, 1966, 820–24.

Beebe, W. *High Jungle.* New York: Duell, Sloan, and Pearce, 1949.

Bourke, J. G. *Scatalogic Rites of All Nations.* Washington: Lowdermilk and Co., 1891; Johnson Reprint Corp., reprinted 1968.

Brown, R., J. Ferguson, M. Lawrence, and D. Lees. *Tracks and Signs of the Birds of Britain and Europe.* London: Christopher Helm, 1987.

Bryant, V. M., and G. Williams-Dean. "The coprolites of man." *Scientific American* 232, 1975, 100–109.

Buck, K. R., P. A. Bolt, and D. L. Garrison. "Phagotrophy and fecal pellet production by an athecate dinoflagellate in Antarctic sea ice." *Mar. Ecol. Progr. Ser.* 60, 1990, 75–84.

Chin, K., et al. "A king-sized theropod coprolite." *Nature* 393, 1998, 680–82.

Coughlan, J., D. J. Bradley, H. Garelick, and D. D. Mara. *Sanitation and Disease: Health Aspects of Excreta and Wastewater Management.* Chichester, England; published for the World Bank by Wiley (New York), 1983.

Decker, D. M., D. B. Ringelberg, and D. C. White. "Lipid components in anal scent sacs of three mongoose species (Helogale parvula, Crossarchus obscurus, Suricatta suricatta)." *J. Chem. Ecol.* 18, 1992, 1511–24.

Denholm-Young, P. A. "Studies of decomposing cattle dung and its associated fauna." D. Phil. thesis. Oxford University, 1978.

Desowitz, R. S. *New Guinea Tapeworms and Jewish Grandmothers: Tales of Parasites and People.* New York: W. W. Norton, 1981.

Devroede, G. "Constipation," in *Gastrointestinal Disease* (eds. M. H. Sleisenger and J. S. Fordtran), vol. 1. Philadelphia and London: W. B. Saunders, 1993, 837–87.

Dix, N. J., and J. Webster. *Fungal Ecology.* London and New York: Chapman and Hall, 1995. (See Chapter 8, pp. 203–24, on coprophilous fungi.)

Dundes, A. *Life Is Like a Chicken Coop Ladder.* New York: Columbia University Press, 1984.

Fry, G. F. "Analysis of prehistoric coprolites from Utah." *University of Utah Anthropological Papers* 97, Salt Lake City, Utah: University of Utah Press, 1977, *xi*–45.

Geneviève (one name only). *Merde!: The Real French You Were Never Taught at School.* New York: Macmillan, 1984.

Gomi, Taro. *Everyone Poops.* Brooklyn, New York: Kane/Miller, 1993.

Good, C. D., J. E. Mars, and E. W. Schmidt. "Waste utilization for propulsion on manned space missions." *Report to Society of Automotive Engineers,* New York, 1968.

Goodrich, B. S., E. R. Hesterman, K. S. Shaw, and R. Mykytowycs. "Identification of some volatile compounds in the odor of fecal pellets of the rabbit, *Oryctolagus cuniculus.*" *J. Chem. Ecol.* 7, 1981, 817–27.

Gootenberg, P. *Between Silver and Guano.* Princeton, NJ: Princeton University Press, 1989.

Haberyan, K. A. "The role of copepod fecal pellets in the deposition of diatoms in Lake Tanganyika." *Limnol. Oceanogr.* 30, 1985, 1010–23.

Halfpenny, J. C., and E. A. Biesiot. *A Field Guide to Mammal Tracking in Western America.* Boulder, CO: Johnson Books, 1986.

Hansard, P., and B. Silver. *What Bird Did That? A Driver's Guide to Some Common Birds of North America.* Berkeley, CA: Ten Speed Press, 1991.

Hanski, I., and Y. Cambefort, eds. *Dung Beetle Ecology.* Princeton, NJ: Princeton University Press, 1991.

Harris, M., and S. Chapman. *Cotswold Privies*. London: Chatto and Windus, 1984.

Heezen, B. C., and C. D. Hollister. *The Face of the Deep*. England: Oxford University Press, 1971.

Hörnicke, H., and G. Björnhag. "Coprophagy and related strategies for digesta utilization," in *Digestive Physiology and Metabolism in Ruminants* (eds. Y. Ruckebusch and P. Thivend). Westport, CT: AVI Publishing, 1979, 707–30.

Hunt, A. P., K. Chin, and M. G. Lockley. "The palaeobiology of vertebrate coprolites," in *The Palaeobiology of Trace Fossils* (ed. S. K. Donovan). Baltimore: Johns Hopkins University Press, 1994, 221–40.

Hurd, P. L., P. J. Weatherhead, and S. B. McRae. "Parental consumption of nestling feces: good food or sound economics?" *Behavioral Ecology* 2, 1991, 69–76.

Hurst, A. F. *Constipation and Allied Intestinal Disorders*. London: Hodder and Stoughton, 1909.

Hutchinson, G. E. "The biogeochemistry of vertebrate excretion." *Bull. Amer. Mus. Natural History, New York* 96, 1950, 1–554.

Jason, L. A., E. S. Zolik, and F. S. Matese. "Prompting dog owners to pick up dog droppings." *Amer. J. Community Psychology* 7, 1979, 339–51.

Kolavic, S. A., et al. "An outbreak of *Shigella dysenteriae* type 2 among laboratory workers due to intentional food contamination." *J. Amer. Medical Association* 278, 1997, 396–98.

Komata, N., et al. "Frass drop samples of beech caterpillars." *Japanese Agric. Res. Quarterly* 28, 1994, 217–33.

Lawrence, M. J., and R. W. Brown. *Mammals of Britain, Their Tracks, Trails and Signs*. London: Blandford Press, 1967.

Lederberg, J. "What the double helix (1953) has

meant for basic biomedical science." *J. Amer. Med. Assn.* 269, 1993, 1981–85.

Leon, M. "Maternal pheromone." *Physiol. Behav.* 13, 1980, 441–53.

Levitt, M. D., J. H. Bond, and D. G. Levitt. "Gastrointestinal gas," in *Physiology of the Gastrointestinal Tract* (ed. L. R. Johnston). New York: Raven Press, 1981, 1301–16.

Lewin, R. A. *The Biology of Algae and Diverse Other Verses.* Pacific Grove, CA: Boxwood Press, 1987.

Macdonald, D. (ed.) *Encyclopedia of Mammals.* Oxford: Equinox, 1984.

Mackay, Charles. *Popular Delusions & the Madness of Crowds.* (Originally published 1841.) New York: Crown, 1995.

Marron, J. A., and C. L. Senn, "Dog feces: a public health and environmental problem." *J. Envir. Health* 376, 1974, 239–43.

Moore, J. G., B. K. Krotoszynski, and H. O'Neill. "Fecal odorgrams." *Dig. Dis. Sci.* 29, 1985, 907–11.

Morris, A. " 'Fight for fertilizer!' Excrement, public health, and mobilization in New China." *J. Unconventional History* 6, 1995, 51–76.

Murie, O. J. *Field Guide to Animal Tracks,* 2d ed. Boston: Houghton-Mifflin, 1982.

NASA, 1984. "Waste collection sub-system study." *Report CR-171836,* National Technical Information Service, Springfield, VA 22161.

Needham, J. *History of Chinese Science.* England: Cambridge University Press, 1978.

NEOS Corporation, "Energy Conversion of Animal Manures." Final report of the Western Regional Biomass Energy Program, Lakewood, Col., 1994.

Noji, T. T., K. W. Estep, F. MacIntyre, and F. Norrbin. "Image analysis of faecal material grazed upon by three species of copepods: evidence for coprorhexy, coprophagy and coprochaly." *J. Mar. Biol. Assn. U.K.* 71, 1991, 465–80.

Patel, P. D., B. F. Picologlou, and P. S. Lykoudis. "Biorheological aspects of colonic activity. 2. Experimental investigation of the rheological behavior of human feces." *Biorheology* 10, 1973, 441–45.

Poinar, H. N., et al. "Molecular coproscopy: dung and diet of the extinct ground sloth *Nothrotheriops shastensis.*" *Science* 281, 1998, 402–6.

Probert, C.S.J., P. M. Emmett, and K. W. Heaton, "Some determinants of whole gut transit-time: a population-based study." *Quart. J. Med.* 88, 1995, 311–15.

Putman, R. J. *Carrion and Dung: The Decomposition of Animal Wastes.* London: Edward Arnold, 1983.

Reyburn, W. *Flushed with Pride: The Story of Thomas Crapper.* Englewood Cliffs, NJ: Prentice-Hall, 1971.

Reynolds, R. *Cleanliness and Godliness, or the Further Metamorphosis.* London: George Allen and Unwin, 1943.

Sabbath, D., and M. Hall. *End Product: The First Taboo.* New York: Urizen Books, 1977.

Sastry, S. D., K. T. Buck, J. Janak, M. Dressler, and G. Preti. "Volatiles emitted by humans," in *Biochemical Applications of Mass Spectrometry* (eds. G. R. Waller and O. C. Dermer), 1st Suppl. Vol. New York: Wiley, 1980, 1085–1129.

Steinbart, H. *Arzt und Patient in der Geschichte, in der Anekdote, im Volksmund.* Stuttgart: Ferdinand Enke, 1970.

Süskind, P. *The Pigeon.* New York: A. A. Knopf, 1988.

Sweeney, W. B., B. Krafte-Jacobs, J. W. Britton, and W. Hansen. "The constipated serviceman: prevalence among deployed U.S. troops." *Military Medicine* 158, 1993, 546–48.

Thacker, E. J., and C. S. Brandt. "Coprophagy in the rabbit." *J. Nutrition* 55, 1955, 375–85.

Thienpont, D., F. Fochette, and O.F.J. Vanparijs. *Diagnosing Helminthiasis Through Coprological Examination.* Beerse, Belgium: Janssen Research Foundation, 1979.

Thulborn, R. A. "Morphology, preservation and palaeobiological significance of dinosaur coprolites." *Palaeogeogr. Palaeoclimatol. Palaeoecol.* 83, 1991, 341–406.

Török, T. J., et al. "A large community outbreak of salmonellosis caused by intentional contamination of restaurant salad bars." *J. Amer. Medical Association* 278, 1997, 389–95.

Turner, J. T., and J. G. Ferrante. "Zooplankton fecal pellets in aquatic ecosystems." *BioScience* 29, 1979, 670–77.

Walker, C. *Signs of the Wild: Field Guide to the Spoor and Signs of the Mammals of Southern Africa.* Cape Town: C. Struik, 1985.

Weisberg, G. P. "Merdre: scatological art." *The Art Journal* 52, 1993, 18–19, 36–40.

White, D. C., D. F. Nivens, A. A. Arrage, B. M. Applegate, S. R. Reardon, and G. S. Sayler. "Protecting drinking water: rapid detection of human fecal contamination, injured, and non-culturable pathogenic microbes in water systems." Proceedings of the North American Water and Environment Congress '96, American Soc. Civil Engineers, New York, NY, 1996.

Wright, L. *Clean and Decent: The Fascinating History of the Bathroom & the Water Closet.* London: Routledge & Kegan Paul, 1960.

INDEX

ABOUT THE AUTHOR

RALPH A. LEWIN is professor of marine biology at the Scripps Institution of Oceanography at the University of California at San Diego. He has also taught at several marine-biology labs, was an instructor and special lecturer at Yale, and has lectured on algae, marine microbiology, and bio-technology in various countries, including England, France, Japan, China, and Brazil. His previous publications, several hundred of them, have been mostly academic and technical. He lives in La Jolla, California.

ABOUT THE TYPE

This book was set in Sabon, a typeface designed by the well-known German typographer Jan Tschichold (1902–74). Sabon's design is based upon thc original letter forms of Claude Garamond and was created specifically to be used for three sources: foundry type for hand composition, Linotype, and Monotype. Tschichold named his typeface for the famous Frankfurt typefounder Jacques Sabon, who died in 1580.

Printed in the United States
by Baker & Taylor Publisher Services